APPLIED ANIMAL FEED
SCIENCE AND TECHNOLOGY

Applied Animal Feed Science and Technology

DR IRVIN MPOFU
[BSc (Hons), MSc, MBA, PhD]

Technical Editor:
MR L. R MUTETWA

Language Editor:
Mr C.K. Vengesai

UPFRONT PUBLISHING
LEICESTERSHIRE

Applied Animal Feed Science and Technology

ISBN 1-84426-271-5

First Published 2004 by
UPFRONT PUBLISHING LTD
Leicestershire

Printed by Lightning Source

PREFACE

This book looks at animal feed science from an applied perspective, with special emphasis on supplementation. Domesticated animals are not free to roam the forests as they feed and this means the artificial environment created for them by man is sometimes (and with some livestock, always) found wanting in as far as supplying all the necessary nutrients. Nutrient supply is critical, especially during periods of rapid growth, pregnancy and lactation. It is, therefore, necessary to feed supplements so as to bridge the nutrient gap created when demand outstrips supply. In commercial animal production, efficiency and effectiveness of feeding are the key elements for enterprise viability. This is achieved through feeding the right feed ingredients, in the right amounts, at the right time and to the right type of animal. It is with this in mind that the author decided to cover pertinent applied animal feed science issues by looking at the nutritional needs of various livestock and compile a dossier on how these classes of animals can be supplemented in order to iron out nutrient deficiencies. The animals covered include beef cattle, dairy cows, sheep, goats, poultry, ostrich, pigs, rabbits, game and horses. The book begins by refreshing the reader on the principles of ruminant and non-ruminant nutritional biochemistry, and then covers the individual livestock classes. The final chapters take a closer look at feed additives, anti-nutritional factors in the form of mycotoxins and biotechnological issues in the animal feed industry. A lot of the examples have been extracted from pamphlets produced by stock feed companies, notably Altech.

The book should, therefore, be appealing to an array of animal scientists, veterinarians, farmers and animal feed science practitioners and technologists. This book is also a useful text for undergraduate university (and college) students.

Postgraduate students will benefit a lot as they can use the book as a quick practical reference book.

CONTENTS

13 General principles of livestock feeding operations

14 Feed formulation techniques

15 Improving the utilisation of livestock feed resources

16 Biotechnology in the feed industry

17 Stock feed manufacturing

18 Establishment of stockfeed manufacturing operations: a case study

The author is grateful to all the peers who made constructive criticisms towards this project. Special thanks go to the technical and language editors, L.R. Mutetwa and C.K. Vengesai.

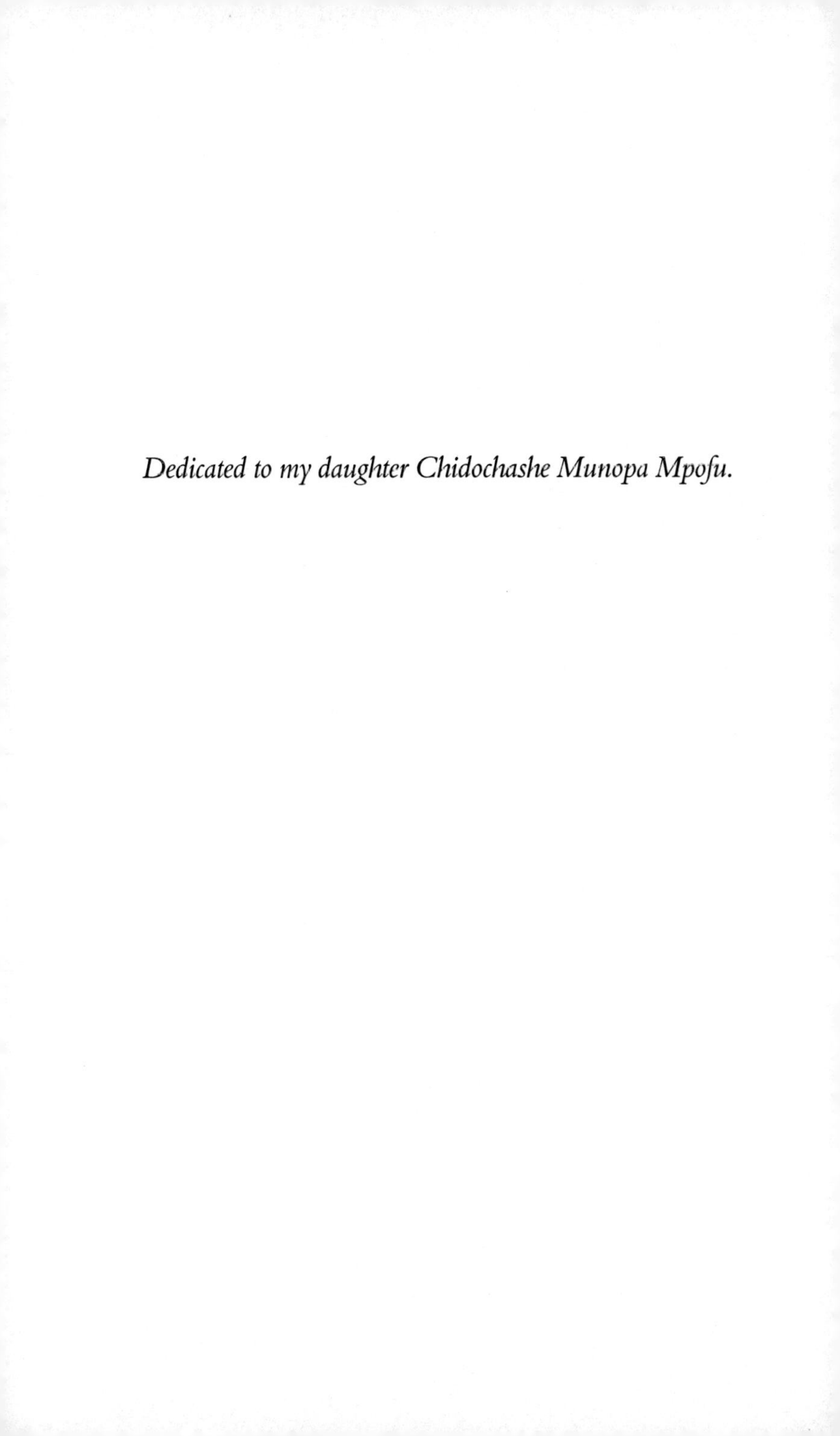

Dedicated to my daughter Chidochashe Munopa Mpofu.

1

BIOCHEMICAL PRINCIPLES OF ANIMAL NUTRITION

1 Introduction

Every animal typically needs a daily supply of roughage (grasses/hay/pastures/crop residues) and growth-promoting feeds (animal and/or plant proteins). For growth and/or maintenance to occur, the animal also requires a daily supply of energy-providing feeds in the form of carbohydrates and fats, and adequate amounts of support nutrients in the form of mineral salts, water and vitamins. Daily provision of these critical nutrients should be balanced so that the animal remains healthy and commands high levels of productivity (reproduction, lactation, daily weight gain, egg production and wool production). The optimum nutritional value is provided for by a situation where animals consume different classes of feeds.

There are four broad classes of feeds, namely roughage (dietary fibre), concentrates, succulents and commercial grade micronutrients (mineral and vitamin premixes) which serve as critical nutrients for metabolism of feed at cellular level.

2 Roughages

Roughages, or dietary fibre, are a class of feeds that consist of indigestible cellulose, hemi-cellulose and lignin complexes of the plant cell walls. The fibre (neutral detergent fibre) content ranges from greater than 50% to 80%. Roughages have a 'scratch effect' and a low water-holding capacity of about 10%, and hence provide bulk (90%) dry matter to the gastro-intestinal tract (GIT). Scratching of sections of the GIT stimulates reflex peristaltic activity and hence aids movement of tract contents towards the posterior and promotes defecation. As a result, roughage or fibrous feeds are important in that they maintain the 'tone' of the GIT, and prevents constipation and other digestive disorders associated with digestion of refined feeds.

The total digestible nutrients (TDN) of roughage range from 40% to 50% with a corresponding energy content variation of 5.8–7.5MJ of metabolisable energy (ME) per kilogram of feed. Roughages also contain proteins, calcium and phosphorus. Protein content varies from 1–2% in old (mature) frosted veld grass to 18% in good quality grass just before full bloom (flowering). Calcium content is high, ranging from 0.5% in green grass to 1.2% in dry mature grass, while phosphorus levels are as low as 0.1–0.3%. Roughages are generally cheap and form the basic fodder in cattle enterprises.

3 Concentrates

Concentrates are feed ingredients that are concentrated in a particular key nutrient. In livestock production, there are basically two classes of concentrates, namely energy and protein concentrates. The nutritional value of concentrates depends upon the composition and concentration of amino acids (in the case of protein concentrates) and carbohydrates (in the case of energy concentrates). It also depends on whether or not the concentrate is easily digestible.

Plant protein concentrates

Vegetable-based feed ingredients generally contain small quantities of protein and the amino acids present are not always balanced to meet animal maintenance, let alone production requirements. As a result, a balanced vegetable protein concentrate normally includes a variety of vegetative materials as a mixture. The only exception to the low protein content of vegetable-based concentrates are legumes. They are commonly used to improve the content of protein in cereal-based diets. For example soya bean meal and sunflower oil meal or cake have protein content values of 41–45% and 42% respectively. Other plant protein concentrates contain 40%, 39%, 32%, 23%, 19%, i.e. for ground nut meal, cotton seed meal, linseed meal, cow pea and jack bean meal, respectively. However, raw soya beans contain anti-nutritional factors, like trypsine inhibitors and urease that interfere with protein digestion specifically in the duodenum and small intestines. Urease can easily cause death, especially in ruminants where it can cause overproduction of ammonia in the rumen. Overproduction of the ammonia gas leads to rumen stasis and alkalosis. Rumen stasis means that the natural rhythmic contraction and relaxation of rumen muscles ceases and this is fatal, as no digesta will be propelled towards the lower gut. Alkalosis means that the blood becomes too alkaline in pH and this offsets various biochemical reactions, especially those dependant on a particular acid-base balance of body cells. Feeding processed soya bean meals that is a by-product of oil pressing normally prevents these problems. The heat generated during oil extraction is enough to denature trypsine inhibitors and urease.

Animal protein concentrate

Most animal-based protein concentrates contain a high concentration of essential amino acids in balanced amounts and are sometimes called first class proteins. Examples of these are fish-meal, meat meal and blood meal. Most protein concentrates contain crude protein content above 30%.

Energy concentrates

Energy concentrates are those feeds that have a high density of easily digestible carbohydrates. Chief examples are molasses and cereal grains. The energy content is very high, above 11MJ/kg of ME with total digestible nutrient level of at least 70%.

Importance of concentrates to livestock

While concentrates are important in ruminants as energy and protein supplements, they are the only ideal staple food for mono-gastric animals. This is because:

- they are low in fibre (<10%); have low moisture content averaging 10%,
- have high levels of phosphorus of 0.4 to 1.2% and
- are rich in digestible nutrients.

Calcium level is, however, as low as 0.05–0.2% and therefore a calcium source is necessary as a supplement to prevent calcium deficiency.

Concentrates are, however, expensive feeds (especially for ruminants) and are only fed as supplements or in diets designed to achieve high levels of production, as is the case with dairy or commercial beef production.

4 Succulents

Succulents are, in most cases, roughage in nature or concentrates when assessed on a dry matter (DM) basis. They are high in moisture content (70–90%) with a DM content of 10–30%. Examples of succulents are lush green veld grass and irrigated pastures. They are highly palatable and, because of their low DM content, are consumed in high quantities of up to three times as much as dry feeds by mass. Animal capacity normally limits amounts consumed. As a result it is recommended that animals on succulent feed (as a basal diet) should receive some daily portion of concentrates if they are expected to produce at a higher level.

5 Support nutrients (vitamins and mineral premixes)

The wide variety of inorganic elements (minerals) that are present in the animal body are obtained on a daily basis from the ingested feeds and, to some extent, from water and soil on feeds consumed by the animal. Minerals are required in many metabolic processes where they maintain the necessary acid base balance but, more importantly, ionic balance for the creation of concentration gradients that are necessary for the transport of solutes across cell membranes. These are also necessary for the proper function of the nerve system. Some mineral forms, for instance limestone flour, are included in feed rations to buffer up the pH of the GIT, especially in the rumen of ruminants. Buffering of rumen pH improves the efficiency of carbohydrate rumen fermentation and protein degradation. The limestone that is used (limestone flour) should be non-caustic and must have the following chemical composition: calcium carbonate 98.24%, aluminium oxide 0.11%, iron oxide 0.07%, silica 1.24%, magnesium carbonate 0.20% and calcium sulphate 0.14%.

Vitamins are long chain organic compounds, present in microscopic quantities in natural feeds, mainly green plants. Although they are organic in nature, they do not possess energy value because the 1 carbon to 2 hydrogen to 1 oxygen ratio does not exist in their molecules. They, however, act as coenzymes, boosting the functionality of natural enzymes so that metabolic activity is always accentuated and not attenuated (impaired) and hence efficiency and effectiveness of feed utilization is effected. Effectiveness refers to the completeness of a reaction (or more of forward reaction in reversible reactions). Efficiency refers to producing more output from less input, or per given level of input.

6 Digestive systems

Providing suitable feeding programs for different species of animals is a big challenge for nutritional biochemists and farmers alike. Many gaps in information have been filled through many years of research. However, continued research efforts are needed, especially on nutrient requirements of different classes of livestock. Planning feeding programmes for the various species of animals requires special knowledge of the different digestive systems. It is the aim of this section to spell out the differences in the digestive tract and make recommendations on the salient facts to be considered in feeding particular animal species.

Digestion in monogastric animals

Digestion in monogastric animals (poultry and pigs) starts in the mouth, commandeered by an enzymatic process closely linked to salivation and mediated by the enzyme amylase (a glycoprotein). The stomach of the monogastric animal is simple and has the following functions:

- regulating passage of digesta by way of pylorus and cardiac sphincters (valves).
- secreting an alkaline, enzyme free, viscous mucus formed by a gel forming glycoprotein that protects the epithelium from acid attack.
- producing hydrochloric acid from oxyntic cells. Hydrochloric acid creates the acidic environment that prevails in the true stomach.

These secretions, in total, form what is known as the gastric juice with a pH of 2 which, in turn, is important in activating protein-digesting enzymes.

Most of the digestion in monogastric animals occurs in the small intestines where digesta is worked upon by enzymatic secretions from the duodenum, liver and pancreas. Because the small intestine is composed of visceral (soft/delicate) tissue, the diet consumed by monogastrics should be very low in fibre (about 3–5%) as compared to diets fed to ruminants and hindgut fermenters (like rabbits, horses and ostriches).

Digestive systems of the hindgut fermenters

This section briefly discusses the digestive system of the ostrich, which is characterised by hindgut fermentation. Hindgut fermenters are animals that digest roughage through a fermentation process, mainly in the caecum. The hindgut is the back-end part of the alimentary canal (composed of the colon, caecum and rectum) as opposed to the foregut, that is the front end, composed of the mouth, esophagus and stomach. As a result, hindgut fermenters are monogastric (simple stomached) animals that exhibit herbivory feeding habits. The ostrich is a good example of a "monogastric" herbivore. The term monogastric has been placed in inverted commas because the system is not so simple as in other monogastrics, like pigs. The ostrich has incredible abilities to utilize roughage. As a result, it is imperative to plan feeding programmes in a manner that is commensurate with the biological and physical capabilities and limitations of the ostrich.

Some important facts to consider are that the ostrich has no (i) teeth, (ii) a crop and that the oesophagus directly empties the food into the proventriculus (sometimes known as the glandular stomach). This means that there is no distinct anatomical demarcation between the oesophagus and the stomach as there is in poultry. The gizzard is found after the glandular stomach, displaying a round and muscular pouch-like structure sometimes called the pebble store. At any one time, a mature ostrich contains about 1.5 kg of pebbles. Their size is just like the size of marbles and they are important in the physical grinding of ingested food due to the rhythmic contraction and relaxation of the gizzard walls. Normally the pebbles are never excreted. They weather or wear away and are naturally replaced by the bird through a natural habit of stone picking. This is why ostriches are seen picking small stones and people think they are burying their heads in sand. In intensive ostrich production systems, it is very important to supply mature birds with marble-sized pebbles all the time. This is, however, not necessary with free ranging birds. Fine gravel and sand should be avoided as the birds can over ingest the material, leading to death. The supply of pebbles has, however, been seen to be more critical when birds are on coarse roughage, or for birds on pasture, than for birds fed correctly processed and balanced commercial feeds.

After the gizzard follows the small intestines, which is about 6 metres long in the adult bird. Then after the small intestines, the ostrich has two large caeca or appendices, closely followed by a very long colon or large intestine. Together these structures are called the hindgut. The colon in large birds is typically 16 metres long. The latter is quite an astonishing feature of the ostrich's digestive tract, but exists for a very crucial purpose. This is what defines an ostrich as a herbivore and a hindgut fermenter as the remains of the plant food leaving the small intestines move slowly through the hindgut and get slowly fermented by friendly bacteria. The fermentation process (anaerobic respiration) changes polysaccharides into mainly volatile fatty acids. The volatile fatty acids are immediately absorbed and utilized as energy sources. Polysaccharides are carbohydrates forming the structure of the plant cell walls. Chief examples are cellulose, hemicellulose and lignin complexes. This basically constitutes what we normally call the fibrous fraction (fibre content) of the plant feed. Some researchers have reported that some amino acids are synthesized in the process that benefits the host animal. These amino acids are mainly due to the friendly bacteria dividing and increasing their numbers, hence their biomass. These bacteria are digested and absorbed in the large intestines by the host animal.

Ostriches practice what is known as coprophagy whereby they re-consume what they have defecated. This ensures maximum utilization of nutrients due to re-digestion and re-fermentation of the herbage.

The overall anatomical features of the ostrich allow it to be able to survive in arid and semi-desert areas, but mainly suit grasslands. As a result, they are natural plant material eaters. They live mainly on succulents (juicy plant materials), seeds, berries, grass, as well as foliage of trees and shrubs. Since calcium is vital in all egg laying species, ostriches naturally supplement dietary calcium by the intake of bones, eggshells and, in coastline areas, seashells. Commercial diets have to address this critical issue, usually in the form of dicalcium phosphate (DCP) (0.53%, –0.72% w/w in starter diets, 1.1% in grower diets and 1.5% in mature maintenance diets). Layers should get up to 3% DCP.

Digestive system of ruminants

Ruminants are animals (sheep, cattle and goats) that have a complex alimentary canal. The complexity is brought about by the fact that instead of a simple stomach, ruminants have a large stomach that is compartmentalised into four chambers. The largest of these chambers is known as the rumen and this is the site for massive fermentation of roughage. Food ingested by ruminants is first soaked and subjected to digestion by enzymes secreted by micro-organisms resident in the rumen, before passing down the rest of the digestive tract. The rumen is a large muscular compartment that is part of the four stomachs that make ruminants peculiar fibre digesters. For effective utilization of feed nutrients in the rumen, a critical phenomenon known as synchronization of energy and protein availability must be achieved at all times. This in turn assures the occurrence of coupled fermentation of proteins and carbohydrates in the rumen.

The synchronization mentioned above ensures the optimisation of the efficiency of utilization of ruminally degraded nitrogen (N) for rumen microbial growth. It is unfortunate that with most natural feedstuffs, this scenario does not always exist throughout the feeding cycle. Fresh grass, for instance, consists of variable amounts of different carbohydrates. Each of these carbohydrates will have different rates of digestion in the rumen such that at some periods within a day, the ratio of available carbohydrates to protein is either greater or less than optimal microbial requirement. This is more important with silage feeding because most water soluble carbohydrates are fermented, significantly reducing the amount of readily available carbohydrates. Ruminants do not require dietary amino acids to the same extent as monogastrics. This is mainly because all the essential amino acids can be synthesized from urea by rumen micro-organisms. Urea is sometimes provided to the animals as ammoniated or urea treated straw. In such a situation, highly fermentable sugars like molasses should be made available to the animal simultaneously in order to achieve coupled fermentation.

The problem in the practical situation is centred on the amounts of urea in the non-protein nitrogen supplements like urea licks or urea treated straw. This is due to the fact that if excess urea is consumed, urease, a rumen enzyme, will produce excessive ammonia gas that in turn causes toxicity (urea toxicity) to the animal. Urea toxicity can kill an animal in a matter of half an hour after consumption. As a result, the level of urea supplements in the diet should be controlled for ruminant diets and fed in conjunction with a readily fermentable sugar like molasses. Urea supplements should never be fed to monogastric animals.

7 Conclusion

On the whole, the difference between digestion of food in monogastrics and ruminants lies in the fact that digestion in monogastrics is mediated through acid hydrolysis in the gastric stomach and by digestive enzymes in the small intestines. Digestion in ruminants occurs chiefly by fermentative metabolism of dietary fibrous components by enzymes produced by micro-organisms resident in the rumen. An important point to note is that hindgut fermentation is important in wild and game animals as an additive digestive process.

Reference

Bellis, DB and R.B Brooks, 'Farm processing of soyabeans for pig feed' in *Rhodesia Agricultural Journal*, 1975,**71**:99.

Journet, M, E Grenet, M.H Farce, M Theriez and C Demarquilly,. 'Recent developments in the nutrition of herbivores', proceedings of the 4th International Symposium on the Nutrition of Herbivores, Clement-Ferrand, September 11 - 15. 1-550 Paris, 1995.

Khalili, H, 'Supplementation of grass hay with molasses in crossbred (Bos Taurus x Bos indicus) non-lactating cows: effects of levels of molasses on feed intake digestion rumen fermentation and rumen digestion pool size', in *Animal Feed Science and Technology,* 1993, **41**:23–38.

Morrison, FB, *Feeds and feeding* 22nd edition, Ithaca, Morrison Publishing Company, 1956, pp.1–90.

Preston, TR, 'Fattening beef cattle on molasses in the tropics', in *World Animal Review*, 1972, **1**:24.

Topps, JH, CA Morgan and J Weir, *Feedstuffs For Ruminants,* Scotland, Scottish Agricultural College, Aberdeen, 1991, pp.1-18.

Topps, JH and J Oliver, 'Animal foods of Central Africa – revised edition', in *Zimbabwe Agricultural Journal, Technical handbook,* Harare, 1993, No.**2**: 8 – 147.

2

BEEF CATTLE SUPPLEMENTARY FEEDING

1 Introduction

Beef cattle, especially in the tropics, subsist largely on veld grazing. Veld grazing in summer is palatable, has enough energy and proteins, but in many areas is low in phosphorus. Bulk is adequate in winter on well-managed properties, but protein content is low. Protein content in summer grazing can be as high as 18%, but drops to 1–2% in winter grazing. Minimum protein levels for maintenance requirements should be at least 7%. Energy content in summer grazing can be as high as 11 MJ ME and drops to 5–7 MJ ME in winter grazing.

The strategic supplementary feeding of protein in the dry season and phosphorus in rainy season to beef cattle is recommended to increase the rate of reproduction and growth.

2 Feeding for better reproduction

Under good management, but with little or no supplementary feeding, an average beef cow will conceive consecutively for two years and then "skip" the next. Accordingly, the cow which has conceived two years running, and the young cow that has to provide for her unborn calf as well as continue her own body development, require a better level of supplementary feeding. In tropical and sub-tropical environments, it is a general recommendation that the producer starts the winter supplementary feeding early in the dry season with about 50% of suggested requirements. This must be increased to about 100% in four months time, and up to 150% to 200% for cows after calving.

Bulls are vital assets on a beef farm and as such they must be maintained in a good and healthy working condition at all times. They should neither be on the thin, light side, nor should they be too fat and heavy.

Where steers are raised, the supplementary feeding level has to be based on the marketing policy of the enterprise and the growth rates expected of them. Ideally, growth rates during pen fattening should be at least 1 kg per day. When pen fattening, the farmer must consider the age, sex, weight, type and value of animals to be pen fed so as to take full advantage of their genetic potential, or to market animals at a younger age, at heavier weights, and at optimum carcass grades. Alternatively the steers can be finished “off veld”. In such a situation, the overall production system should be designed for a cattle growth pattern such that only a small daily supplement is required in the final finishing phase. A fattening programme that involves buying and selling the fattened stock can be tricky because it is easy to make a financial loss, even before starting, by paying too much for the animal. This can be avoided by doing routine budget projections, so that purchases guarantee profits if the efficiency of fattening is upheld.

3 Veld fattening beef cattle on grass

Veld fattening beef cattle on grass could be very profitable as it involves little or no purchase of concentrates. If it becomes necessary, the use of concentrates should be carefully balanced when used with green grass to maintain efficiency. This is vital in situations when the particular beef cattle do not adapt to pen feeding, and in systems where there is abundant grazing, but limited grain is produced, or on good, irrigated pastures. If the whole aim is to achieve pre-fattening target mass, it is necessary for animals to grow as much as possible when they are weaners and yearlings, and to lose little or no mass during winter while green grass is an effective nutrient source for carcass gain. Farmers must, however, prevent winter loss by being prepared to provide supplementary feed.

Average growth responses follow three distinct phases relating to live mass and carcass growth as shown below:

a) Early summer: live mass gain/animal/day = 0.5–0.6 kg
b) Midsummer: live mass gain/animal/day =0.7–0.8 kg
c) On irrigated, heavily fertilised pastures, live mass gain/animal/day = 0.8 kg

The farmer must beware of over feeding. Large quantities of concentrates can have a substitution effect on grass intake. The amount necessary should be fed judiciously.

As mentioned before, animals should be selected strictly on pre-fattening target mass to ensure top grades. As a matter of fact, animals younger than two years sometimes have difficulty in consuming their ration of concentrates, and are unlikely to reach their target mass within the specified time period. On the other hand, animals close to 400 kg at 18 months of age may succeed with grass fattening provided the grass is palatable and of high quality, e.g. Katambora Rhodes grass, Star grass or Midmar rye grass. It is also important to note that breeds differ widely with respect to the mass at which they attain slaughter grade and that slaughter mass is also influenced by age.

Overall, to achieve good results, it is necessary to provide adequate protein and energy. Most veld fattening cubes are specially formulated for this purpose. These are also available with an ionophore inclusion. Ionophores help with active uptake of calcium and phosphorus from the digestive tract into the blood circulatory system.

Table 2.1 overleaf shows the various types of supplementary feeds that farmers can adopt for profitable beef production during the rainy and dry season under tropical conditions.

Table 2.1 Supplements for the rainy and dry season

Season	*Type of supplement*
Summer (rainy season)	Mineral mixes containing 7% phosphorus and vitamins A and E.
Winter (dry season)	(a) Protein containing meals (commercial grade), i.e. fattening meals. (b) Farm gate mixtures of 3 parts maize meal to 2 parts soya bean meal. Can also substitute portions of soya bean with sunflower meal, groundnut meal, groundnut tops, wild pods, fish meal, meat and bone meal, blood meal, cotton seed cake (up to 5%) and maize stover or veld hay treated with urea. (c) Feed conversion efficiency enhancer, e.g. ionophores, Browse Plus and fibrozyme. Feeding rate range of these enhancers is between 10–15 g per animal per day.

4 Pen fattening cattle on a standard high energy diet

Pen fattening is a process of feeding animals in pens so that they literally put on some intramuscular fat and reasonable subcutaneous fat. During the fattening process, the animal gains weight and its leanness improves also. The animals are fed in pens so that movements are limited, hence they do not waste energy (and body reserves) by moving about looking for food. In other words, pen fattening means feeding animals in feed pens or troughs (zero grazing).

Pen feeding aims to:

- Fully express the animal's genetic potential for growth and, therefore,

- Market younger, heavier animals at optimum grades (super or second grade).
- Maximise profit per animal, by pen feeding when the beef price to feed cost ratio is favourable.

Pen feeding should begin with a series of adaptation periods namely:

- 0–3 days add 30% milled roughage
- 4–6 days add 20% milled roughage
- 7–9 days add 10% milled roughage
- 10+ days final mix ad-lib to finish

Then a quick budgeting exercise is done to assess the induction, target mass, and general performance on the high energy fattening diet.

5 Fattening meals

A typical pen feeding meal has around 12% crude protein and less than 2% urea. This is a fully balanced meal that must be fed ad libitum (without any restrictions).

Another option involves mixing 9 parts pen meal to 1 part milled hay. In this case the pen meal has 13.5% crude protein and urea content of 2%. This should not be fed without the aforesaid mixing otherwise urea poisoning may occur. The third option involves using a beef fattener concentrate which has 17% protein and 3.6% urea. This concentrate must be mixed with snap corn or its equivalent on a 1:1 weight basis.

The fourth option involves a mixture that is very popular with many beef fatteners in Southern Africa. The concentrate is mixed 1 part of a beef concentrate to 4 parts of snap corn. This concentrate contains 9 % urea and the crude protein content is 32%. If the urea content is increased to 12%, we have option 5 that has 64% crude protein with 12% urea and must be mixed with snap corn or its equivalent on a 1:9 weight by weight basis.

In order to improve feed conversion efficiency an ionophore can be included. Most ionophores have additional advantages in that they are coccidiostatic. Use of ionophores should adhere to manufacturer's instructions normally printed on the package.

6 Feeding period

Animals on a low protein and energy concentrate are normally fattened for 130–150 days while those on a high protein and energy concentrate can stay in the pens for as little as 70–90 days. However, for practical purposes, the feeding period should be long enough to enable the animal to reach optimum potential of fleshing and fat cover.

As a result, it is imperative to monitor progress by weighing a representative sample. This allows the producer to accurately record the feed intake and relate it to daily weight gain. Experienced farmers normally sample well-fleshed animals for slaughter from the fattening herd by merely looking at the animals. This practice should not, in any way, replace the sample weighing completely. Sample weighing allows the calculation of marginal efficiency, that is basically cost of beef per kg divided by the cost of feed per kg consumed during the period of pen feeding.

The main factors affecting profitability of a pen-feeding programme include:

- Cost of animal (per kg cold dressed mass),
- Cost of the feed per kg,
- Feeding period and
- Efficiency of feed utilization.

The efficiency of feed utilization should be about 8 kg dry matter of feed for every 1 kg of meat (on live mass basis). Health status of the animals is also critical, especially control of stress-related diseases such as bovine respiratory disease-complex.

7 Strategic supplementary feeding for grazing beef animals

Tropical grasslands generally have two types of veld, namely the summer veld and the winter veld. The summer veld grazing is palatable. However, it is, in many areas, low in the element phosphorus. Phosphorus is important in maintaining the integrity of the skeletal tissues and in the efficient utilization of energy from feeds during protein, fat and carbohydrate metabolism and in the proper function of many enzymes. As a result, phosphorus can increase calving percentage by 5%, weaning weight by 10 kg and live mass gain by up to 40 kg. These responses, however, vary from one agro-ecological locality to another. For example, in an agro-ecological area where there is abundant good quality veld pastures during the summer months, the benefits of phosphorus supplementation are on the higher end of the performance figures given above. In agro-ecological regions with satisfactory quality veld pastures, above average performance is obtained from phosphorus supplementation during the peak rain season when veld grass is adequate. When feed scarcity sets in as the dry season ensues, the benefits are on the lower end of the scale. But if the feed scarcity is corrected through supplementary feeding with good quality roughage and proteinous sources (commercial or farm gate), phosphorus supplementation improves calving percentage, weaning weight and live mass gain remarkably.

Phosphorus supplementation of beef cattle

Phosphorus (P) is one of the most important mineral elements in animal nutrition. It is, therefore, important to devote a section on this mineral for the benefit of the beef producer and his/her animals. Together with calcium, phosphorus is vitally important for the development of strong teeth and bones and is, therefore, essential for growth and reproduction. It is also critical in all energy biochemical reactions in the animal tissues.

Many soils in the tropics are deficient in phosphates. Cattle producers should consider carefully the soil type in their area. A better understanding of why, when and how phosphorus should be fed will lead to increased productivity.

Phosphorous deficiency and its effect

Symptoms of phosphorus deficiency include lack of appetite and a rundown, uneconomic appearance. Growing animals fail to make normal gains, while in lactating cows, milk yield will drop significantly. A continued lack of phosphorus will cause a depraved appetite which leads to poisoning (botulism) due to the consumption of decayed bones or, in some cases, flesh on bone surfaces.

Animals deficient in phosphorus over a long period will develop stiffness of the joints that may become fragile and brittle. High producers and young stock are affected more by the deficiency because of their greater demand for the mineral.

Breeding cows fail to come on heat regularly. If and when they do, the body of the mother does its best to protect the young against the deficiency by withdrawing the mineral from her skeleton. As a result the cow is weakened, her productivity is reduced and she will raise a poor calf.

Factors to consider when supplementing phosphorous

Two factors are of vital importance when considering phosphorus supplementation. These are:

- The percentage of phosphorus in the grass, and
- Stage of life cycle of the animal.

This could be in terms of growth, lactation or maintenance.

When grass matures, the percentage of phosphorus decreases, but the content is generally sufficient for maintenance levels until late maturity if fed in conjunction with winter maintenance feed. Once the rains have fallen and green grass starts to grow vigorously, the phosphorus levels also increase. The needs of the animal also increase, due to growth and increased body weight. Thus, it is generally accepted that the best response to phosphorus supplementation occurs soon after the onset of the rains.

However, a phosphorus requirement increases with the animal's energy intake, as it is a critical element in the metabolism of dietary energy. As energy intake may be dependent on season, this is an important factor and is especially relevant in breeding animals.

Table 2.2 gives the requirements for phosphorus of growing steers (ARC 1980).

Table 2.2 Phosphorus requirements in growing animals

	Daily Gains – g/day				
	0	250	500	1000	1500
Live mass (kg)	*Phosphorus requirement (grams/day)*				
200	3.1	5.4	8.3	13	14
300	4.6	6.9	9.3	14	19
400	8.3	11.0	15	21	27

Once the steers are gaining weight in summer their phosphorus requirements can be 3 to 5 times the level of that at maintenance, depending on their live mass. The green grass contains 0.14% phosphorus in the dry matter and, with a dry matter intake of 5–10 kg/day, the green grass will only provide up to14 grams of phosphorus per day.

Of major importance to cattle producers is the phosphorus requirements of pregnant cows. Table 2.3 gives these requirements (ARC 1980).

Table 2.3 Phosphorus requirement of pregnant cow

	Week of pregnancy					
	20	24	28	32	36	40
	Phosphorus requirements (g/day)					
Calf 1	11	11	12	20	22	26
Calf 2	13	14	15	17	19	22

These figures should be used when planning for the next calving season. In the majority of tropical areas, this happens when the rainfall is mono-modal prior to the rains, or at about the onset of the wet season, i.e. October to December in Southern Africa.

The phosphorus status of veld grass during winter is very low at 0.09%. While 0.09% phosphorus combined with a protein supplement would be sufficient for maintenance, this is not adequate for pregnancy or lactation. Increasing the protein value of feed would only help slightly, as the grass intake is not increasing and, therefore, phosphorus supplementation is also required.

Phosphorus supplementation is of major significance in first calving heifers. Their requirements are high and the animal is still maturing. This is because the young cow has a calf at foot drawing milk and it is still expected to conceive again.

Benefits that are gained by supplementation of phosphorus in beef production are:

- Improved fertility,
- Improved live weight gains in breeding females and yearlings, and improved weaning weights.

Phosphorus supplemented cows do not lose weight at the end of the rainy season, but continue to improve, thereby ending the season in a far better condition than those not supplemented. This, of course, assumes that the other chief nutrients (proteins, energy) are not limiting.

How to feed phosphorus to beef cattle

Phosphorus feeding strategies require that initially the animals be fed on an ad lib basis. This usually results in high intake. As the season progresses and the animal requirements are met, the intake decreases until late summer when the animals will be taking a nominal amount.

The palatability of the ration is important. It is possible to mistake low intake for satisfied phosphorus levels rather than a dislike for the product by the animals. This, therefore, should be a deciding factor when choosing a ration.

Commercial rations come in both lick (loose form) and block form. The latter is easy to feed, but intake can vary depending on the block's hardness, while the lick gives more consistent intake, but needs specially modified containers (i.e. used tyres or half drums) to store it in when feeding.

Phosphorus requirements for breeding cows and growing steers

Breeding cows require 10 g of phosphorus per cow per day and should receive, on average, 145 g of a 7% phosphorus block or lick. If a 4% phosphorus block or lick is used then to meet the 10 g daily requirement, each animal should consume 250 g per day of the supplement. Growing steers require less phosphorus requirements per day per animal, i.e. 5 g. Thus, if a 7% phosphorus supplement is used, the animal should consume 75 g and 125 g if the phosphorus block or lick contains 4% phosphorus. The number of phosphorous blocks used can easily be calculated depending on the size of animal groups. Normally, 15 kg phosphorous (7% P) block is adequate for 20–22 beef steers cattle per day.

Protein supplementation for beef cattle

The winter veld bulk may be adequate on many farms, but the protein level is always low, i.e. below 7%. The protein level in the grass declines as the winter season (dry season) progresses to as low as 2%. Protein supplementation is imperative, not the exception in order to prevent weight loss, or at least reduce the extent of weight loss. Protein supplements always improve dry matter intake and digestibility. Intake can increase by as much as 15–45% while digestibility can increase by a margin of 2–5%. Normally when such responses are obtained, protein supplementation is more cost effective than the use of energy supplements. It is generally suggested that protein supplementation starts long before summer gains cease. This calls for sample weighing (10% of the herd) fortnightly. The aim is to assess the trend as regards weight gains so that the timing of supplementation does not start when there is a problem already. Protein concentrates come in cube or block form of 30 to 45% crude protein content. As a general guide, a 30% crude protein block should be fed to pregnant heifers and immature cows at a rate of 800 g per day per heifer for the animals to receive the critical requirement of 240 g protein per day per animal. To achieve the same requirement, a 35% protein cube should be fed at a rate of 685 g and a 45% protein block or cube at a rate of 535 g per day per animal.

Mature cows require 145 g of protein per day per animal. Protein supplementation should, therefore, be 483 g with a 30% protein block, 415 g with 35% cubes and 322 g with a 45% block or cubes, per day per animal. Yearling heifers require 180 g of protein per day per animal so they must consume 600 g of the 30% protein block, 514 g of the 35% protein cube or 400 g of the 45% protein block or cubes. Two and a half year old bulling heifers require 205 g protein per day per animal. As a result, the feeding options are 683 g if a 30% protein block is used, 585 g if it is the 35% protein cubes and 455 g if it is the 45% protein block or cubes.

All in all, supplementation should preferentially start with pregnant heifers and immature cows, followed by pregnant weaner heifers, yearling heifers and lastly the two-and-a-half-year-old bulling heifers.

For effective results from protein supplementation, the cows should be grouped by supplementation needs into, say, herds of heifers, herds of first calf cows, second calf cows, mature cows and bulls. It is important to underline again that the farmer should not wait until the cattle are thin to start supplementation. If this happens more often than not, the animals will require twice as much supplements per day per animal and it is obviously not economic.

Feeding the higher levels of supplement requires balancing for micronutrients. Trace minerals must be added to the supplement, especially if mineral supplementation is free choice, because the intake of the latter normally declines as the animal spends more time on the concentrates.

Towards the end of the dry season, beef animals rely on foliage from trees and these are high in anti-nutritional factors like tannins. Additives in the form of *Browse Plus* have been used with varying successes in Zimbabwe. Generally speaking, *Browse Plus* is a rumen modifier that can be added to the feed to improve rumen function, particularly the availability of crude protein from plant material high in tannin and lignin. Tannins and lignin bind to proteins and this process renders them undegradable in the rumen and hence deprive the animal of the much needed nutrient. Best results with *Browse Plus* have been found in late winter or during droughts when browsing is an important feeding behaviour or pattern. It can also be incorporated into the 30% crude protein block, 35% crude protein block and 45% crude protein block or licks.

Bull feeds

Bull testing cubes ($\pm$12% CP) are specifically formulated to evaluate the performance of young bulls under the Bull Performance Scheme. This complete feed contains appreciable amounts of roughage and should be fed on an ad lib basis. A concentrate meant to condition the bull that has passed the performance test should contain a higher crude protein content of 16%. If fed to working bulls running with the cows, this feed supplements natural protein and helps to maintain condition. Fed at 0.5–1% of body weight, these cubes can also be used prior to the bulling season to ensure bulls start in optimum condition.

Beef quality and feeding

Beef quality is important to the consumer, especially in highly competitive markets. It has become more of a requirement to feed vitamin E so that the meat quality improves.

The major contributing factor to meat quality deterioration is lipid oxidation. Vitamin E is a fat-soluble nutrient that is essential for growth and health in animals. It happens to be the body's most potent antioxidant. It is located in the lipid bi-layer of cell membranes and it protects cells and tissues from oxidation and free radical damage.

Appearance and drip loss

When a customer shops for meat in the butchery or supermarket, he or she expects, or should be able, to see at a glance that the meat is good quality. Good quality meat has a clean, dry appearance when packaged. This means it should display less drip, meaning there is less weight loss on the shelf. Reduction due to drip loss can be as high as by a factor of 30–40%.

Feeding vitamin E to beef cattle reduces drip loss in both fresh and frozen meat. Vitamin E also extends the shelf life of the meat and reduces the need for artificial antioxidants or any other commercial substances that are normally used to prolong shelf life. Feeding adequate vitamin E has an added advantage of maintaining the actual colour of the meat for as long as over 12 days of retail display. Colour intensity and consistency are also maintained.

Lipid oxidation is a major factor contributing to meat quality deterioration. This phenomenon occurs in all common types of meat (poultry, beef and pork). Lipid oxidation affects shelf life, colour, appearance and sensory properties of meat. These are all-important factors for consumers.

Lipid oxidation begins in meat immediately post-mortem. For example, the off-flavours noticed in meat after a few days storage are due to post-mortem oxidation as is the brown colour that develops in red meat. However, meat can be protected from oxidative deterioration. Nature has provided meat with built-in antioxidant protection. This protection is provided by vitamin E which, in its alpha-tocopherol form, protects cell membranes against oxidative damage. It is now established that feeding animals more vitamin E delays post-mortem oxidation. Thus, meat and meat products have inherently better quality factors for shelf life, colour, appearance and sensory properties are improved. In supermarkets and other retail outlets, the meat on display plays a key role in shaping customer perceptions about the entire store and influences consumer buying habits.

8 Conclusion

This chapter has highlighted the fact that supplementation of some sort is needed in all seasons for beef animals. During the rainy season (summer months), when there is generally abundant lush green grass, protein and energy requirements of grazing beef animals are easily met. But because the utilization of nutrients largely depends on the supply of phosphorus, the latter is found to be in short supply during this period. So, during the rainy season, beef animals should receive phosphorus supplementation. During the winter and spring months (dry season), beef cattle are always in short supply of proteins. As a result, during the dry season, the chief nutrient to be supplemented to beef cattle are protein concentrates of various forms, i.e. bought in commercial grade or farm gate preparations.

References

Agricultural Research Council. 'The Nutrient Requirement of Ruminant Livestock', in *Commonwealth Agricultural Bureau*, UK, 1980.

Cattle Producers Association, 'Beef: Feedlotting Beef Cattle', *Beef Production Manual*, Commercial Farmers Union, Zimbabwe, 1998, No. 98/8, pp.1-22.

Gammon, D.M, 'Fattening rations for beef cattle', in *Beef in Natal*. Department of Agricultural Development, Pietermaritzburg, 1992, pp.1-15.

McDonald, P, R.A Edwards and J.F.D Greenhalgh, *Animal Nutrition*, fifth edition. Harlow Longman, 1995.

N.R.C, *Nutrient Requirements of Beef Cattle,* revised seventh edition, Washington, D.C, National Academy Press, 2000.

Wiseman, J. and D.J.A Cole, eds., *Feed Evaluation*, London, Butterworths, 1990, pp.1–28.

3

DAIRY COW NUTRITIONAL CHALLENGES

1 Introduction

Many farmers and dairy scientists have described a dairy cow as a factory that transforms feeds unusable by humans into milk. The feeds the cow eats, in addition to producing milk, must also supply nutrients used for growth of the cow's body, reproducing calves, and for the provision of nutrient stores, better known as body reserves. These body reserves are vital in that they are later utilised by the cow during times of deficit feeding, thus when requirements are greater than intake. Different levels of nutrients are required, commensurate with the dictates of a particular phase of the lactation cycle. This chapter seeks to explore these situational requirements and recommend supplementation strategies for optimum dairy production.

2 Lactation cycle and nutritional challenges for dairy cows

During early lactation, milk production increases rapidly. Feed intake, however, does not keep pace with nutrient needs for milk production, particularly energy, and body reserves must, therefore, be utilised. If nutrient requirements are not met, low peak production levels result, and this then results in an overall low lactation yield. Protein content is critical during this period, and requirements vary depending on the production levels of the cows.

The shorter the length of time the cow is in a negative energy balance (i.e. she is unable to source sufficient energy from the feed alone, and must therefore utilise body reserves), the quicker feed intake can be maximised and milk yield increased.

Feed intake reaches a maximum anywhere from 10 weeks after calving down. This is also when lactation peak occurs. Peak milk production can, therefore, be maintained by maximising on nutrition at this stage. Feed quality and nutrient availability play a very significant role in ensuring optimal production.

In mid-lactation (from 140 days), milk production starts to decline significantly and progressively. Milking cows should at least be maintaining body weight at this point, if not gaining slightly. This is the best time to start feeding for condition in order to have the cows drying off at the correct body scores. Nutrient requirements towards the end of lactation are more easily met, and body condition can be improved. Since lactating cows require less feed to replace one kg of body tissue than dry cows, it is more efficient to have the cows gain condition near the end of lactation, rather than during the dry period. Young cows (heifers) and first and second calving cows should be fed additional nutrients to allow for their own growth as well as for milk production.

Challenge feeding involves presenting the cow with more than she normally eats in order to encourage intake and thus reach peak milk production at an earlier stage, which translates into higher overall milk yields.

3 Home mixing and the dairy farmer

With the introduction of mechanical feed mixers in most countries, more producers are moving into complete home mixing systems. But whether you have a mixer or not, it is still necessary to ensure that the correct levels of minerals and vitamins are being made available to your cows.

Mechanical mixers have the main advantage of eliminating any selective preference the cow may have regarding feed intake. Concentrates are often eaten in preference to the roughage, and this creates an imbalance in the rumen, as far as pH and bacterial population and functioning are concerned. The uniform blend of ingredients obtained from a correctly functioning mixer ensures that the same feed is being offered at every feeding episode. Feed mixing on the farm should be scheduled on a particular day of the week, say every Tuesday. So what this means is that Tuesday should largely be set as a day when feed for the next two weeks is mixed such that at any one time there are food reserves for a week in case of unforeseen circumstances, like late procurement of ingredients or machine breakdowns. Daily mixing should only be done if ingredient supply, power and workability of the mixing machine are absolutely guaranteed.

The correct way of mixing a ration involves:

- Correctly weighing the feed ingredients,
- Putting, the bulk portions (i.e. roughage and concentrates) first and
- Blend them thoroughly,

- Then those ingredients that are to be included in smaller quantities (i.e. vitamin, limestone flour, common salt and minerals) are added last and
- Finally blending the whole lot of the ingredients together.

Normally, experienced workers can tell when the mixing is perfect by merely looking at the colour changes during mixing, otherwise the recommended mixing period should be followed, normally 20–30 minutes, depending on the machine's rpm (revolutions per minute) rating.

The mineral and vitamin premixes available from most suppliers are specifically designed to provide all the necessary minerals and vitamins that the producing cow requires for optimal milk production and regaining of condition. One unit is normally added to one tonne of feed, but it is advisable to follow the manufacturer's instructions usually printed on packaging.

A drive towards home mixing

Home mixing involves purchasing some feed ingredients and the growing of some raw feed materials so that rations are made on-farm. Farmers who mix maize with concentrates that are bought in are only partial home-mixers.

There are several reasons why many producers the world over are now home mixing dairy feeds. These include:

- High stock feed costs,
- Increased availability of raw materials on-farm,
- Greater access to feed ingredients/materials from stock feed companies.

Costs that must be considered when evaluating the benefits of a home-mixing programme include

- The existing and future storage capacities and facilities,
- The transport of any raw materials/ingredients bought,
- Additional labour costs,
- Interest paid on stock held,

- Equipment needed to facilitate mixing, and
- Weighing animals for monitoring purposes.

Stock feed costs have risen as a result of:

- Higher prices being demanded by the producers of raw materials,
- High overheads,
- Increased power and water charges,
- Limited availability of cheaper by-products,
- Higher transport costs and
- Labour costs.

Most home mixers will appreciate these costs, as they constitute a major part of any feed producing enterprise, regardless of size of output. In fact, in Southern Africa the large margins which were previously characteristic of any home mixing system versus the use of stock feeds diminished significantly over the past few years in real dollar terms, due to the increased costs of production facing the farm mixer.

Aside from these important factors, the question of correct nutrition as far as ration formulation is concerned must be addressed. The following questions come to mind:

- *Do the farmers know how much protein, energy, fibre, mineral and vitamins the ration they intend to use provides, and how much their cows need?*
- *Are the farmers including a comprehensive mineral/vitamin premix in their rations?*
- *If they are, do they know what goes into it, and how it compares with recommended levels?*

Using the correct mineral and vitamin premix is vital for optimal metabolic functioning, and can have limiting effects if levels are not sufficient. Feed will be utilised less efficiently than otherwise, such that the cows will produce less milk per kilogram of feed eaten. This, in turn, makes the whole operation expensive.

Premix manufacturing companies follow guidelines of recommended levels provided by the international research councils like ARC, NRC, INRA and from local research centres. Most feed manufacturers execute their own validation research programmes.

4 Dairy feeds

Strong healthy calves are vital to the dairy herd and should be cared for accordingly. They are a source of genetic improvement and represent a farmer's future milking stock. Top quality feed, combined with good management, will result in a strong foundation for continued success. The feeds must be balanced for all the key nutrients. The section that follows briefly covers the feed supplements that can make dairy production profitable.

Calf meals

The farmer can choose between a 20% crude protein (CP) or 16% CP level supplement, depending on cost as opposed to the growth rate envisaged.

Calf rumen starter

This is normally used for the Johnson calf rearing method. It is advisable to feed calf rumen starter first, ad lib (without any restrictions, meaning the calf should be allowed to eat as much as it can) until weaning from milk.

Calf meal

The 16% CP calf meal should be fed ad lib from day one to six months of age, where the Johnson calf rearing method is used. Good roughage should gradually be introduced separately from three to four months of age. The good quality roughage stimulates and primes the calf to start ruminating (refer to chapter 1). After six months of age, the 16% calf meal should be rationed at one to two kilograms per calf per day. Thereafter, the calf meal (20% CP) can be offered at a rate of 1 kg per calf per day. Alternatively, the farmer can opt for the 16% CP calf meal.

Bovatec supplement

Bovatec supplement, or any equivalent, is used as a carrier for the ionophore Bovatec, to improve growth rates of heifers if fed from six months of age to calving. Normally the concentrate being fed is reduced by one kilogram and replaced with one kilogram of Bovatec supplement.

Bovatec supplement may also be fed when summer fattening on grass is an option. The feeding regime is outlined as above, i.e. one kilogram per head per day.

Supplements for lactating cows

Supplements for high producing cows come in the form of a meal or as cubes. These are physical forms meant to reduce feed dust and hence improve feed intake. High producing dairy cows produce anything between 25–35 kg of milk per cow per day (some can be conditioned through genetics (e.g. Holstein breed) and feeding to produce 40 kg per day or more). The dairy meal supplement should therefore contain adequate protein levels of between 18% and 21%. The 20% dairy meal is popular in many countries.

Super dairy meal/cube

This is a high energy feed commonly known as super dairy meal and is supposed to give super results as the name goes. It is ideal for top producing cows. Feeding is recommended at 400–500 grams per kilogram of milk in conjunction with good quality grazing and hay. This is sometimes referred to as challenge feeding. In other words, the animal is supplemented on the basis of the kilograms of milk produced per day, and each time the daily milk production changes, upwards or downwards, the supplement offered is adjusted accordingly.

Cow premix, Ca – 18.6%; P – 6.3%; Salt – 20%

The cow premix is essentially a vitamin/mineral mix supplement that contains 18.6% calcium, 6.3% phosphorus and 20% common salt. The remaining portion is composed of an assortment of other key minerals and vitamins that are important in promoting high milk production. Common salt is an important appetite inducement. The cow premix is for home mixing with farm grown maize and oil cakes. This premix supplies the complete mineral/vitamin requirements when included at 5% of the mix. The cow mix should ideally be incorporated in the final diet in bulk. So for every tonne of the diet, 50 kg of cow premix is required.

5 Dairy feeding and viability

When it comes to viability, feed is the single biggest factor in the equation. Dairy producers know that only by feeding the best quality feed will their cows achieve the desired results, at cost-competitive rates.

For example, reproductive performance is severely curtailed by poor nutrition and the breakdown of financial losses post calving of cows over 90 days is normally recorded at around the following levels (table 3.1):

Table 3.1 Financial losses due to poor nutrition in a dairy herd

Loss of milk production	*32 %*
Added veterinary and medical costs	*13 %*
Calf loss	*13 %*
Added breeding costs	*11 %*
Replacement costs	*31 %*

Production demands and stresses increase the incidence of reproductive failure. Many of the reproductive problems in dairy cows are related to mineral deficiencies of copper, zinc, manganese, or selenium. Examples of reproductive failures include cystic ovaries, embryo re-absorption, silent heats, abortions, poor uterine tone, endometritis and poor sperm formation.

Trace minerals influence hormonal patterns, integrity of epithelial tissue and the immune system. The inclusion of Bioplexes in dairy diets (organic forms of copper, selenium, manganese, and zinc) during the first 100 days following parturition has been shown to increase fertility in dairy cows through increased conception rate and improved embryonic survival. Bioplexes are biological complexes formed by conjugating an inorganic mineral with an organic compound like an amino acid. The complex makes the mineral more bioavailable so less minerals can be included in the diet than with inorganic sources.

Research in the 1990s has shown unquestionable benefits of quality feeding by including Bioplexes in dairy cow diets.

For example, Wiersma of the Animal Nutrition Department of Central Wisconsin (USA), showed that the addition of bioplex copper, zinc and manganese resulted in fewer services per conception and improved conception rate in dairy cows. Fallon at Teagasc, Grange Research Centre in Ireland, showed that Bioplex supplementation favourably affected fertilization rate and number of fertilised embryos in superovulated heifers. Researchers at the University of Dublin found that Bioplex copper, zinc and selenium reduced the number of days to first follicle production and number of days to first service.

A study performed at a bull station in the Midwestern United States showed Bioplexes improved viable sperm counts, transmittance and number of units of semen collected. This is not surprising since zinc is required for spermatogenesis and seminal fluid production. Bioplexes are normally incorporated in feeds, liquid supplements, and mineral blocks.

While most of these studies were done in temperate environments they are still applicable to tropical dairy cows because the biology of the respective cows is still the same, nutrition-wise.

6 Organic minerals for dairy cows

Other additives and premixes that are key to improving the quality of feeding dairy cows also comes in a combination of 4 organic minerals, zinc, selenium, copper and manganese, and play an important role in continued reproductive functioning. For example, manganese is necessary for normal oestrus and ovulation, while zinc is vital in spermatogenesis and maintenance of epithelium in the genital tracts. Copper deficiencies may show early embryonic death and rapture failure. Selenium supplementation results in improved viable effective production and thus lowered number of inseminations needed for a successful conception.

Generally, it has been seen that the use of bioplex minerals (Zn, Cu, Mn and Se) helps in improving conception rate and results in fewer number of services per conception.

As a result, most of the minerals known to be of particular relevance to milk production in dairy cows are available in an organic bioplex form. This form of mineral presentation maximises, updates and is used in all stages of growth, maintenance, reproduction and lactation of today's dairy cow.

Zinc is of particular importance in lactating cows as it is a vital component of the keratin lining that is shed from the teat at every milking. This lining constitutes the first line of defence against microbial attack, and poor regeneration will result in infection and subsequent mastitis. Keratin is also important in hoof strength and integrity.

The Bioplex packs should be added to a complete feed mixture at a rate of 1 pack (200 g) per tonne mixed feed. These packs should be used when general requirements are believed to be insufficient, or infertility is evidenced.

7 Winter feeding of dairy cows

It is a well known fact that during winter, and even in the poorer summer seasons, the quality of forage available to the producing animals is short of the necessary nutrients required for optimum metabolic functioning. Energy levels in particular are often low and it is for this reason that a high-energy ration is designed to accommodate the fluctuating forage values and additional energy requirements.

The need for high energy rations occurs in those animals that are in early lactation, especially first calvers that are still growing, animals in areas with poor quality forages and in high producing cows.

Various ranges of high energy feed have been devised for effective feeding under these listed conditions. The high energy range of feeds provides an energy rich feed source, while ensuring that these listed conditions correspond to protein levels and are sufficient. This is particularly important because without enough protein, the energy use is limited, and production is not optimal. When comparing prices of feeds, it is important to consider both the protein and energy levels before deciding on which feed is best suited to animal requirements. Examples of the high energy range of feeds are given in table 3.2.

Table 3.2 Examples of commercial high energy feeds used in winter feeding

	CP%	MJ/KG
High energy 160	16	11.7
High energy 180	18	11.9
High energy 220	22	12.0

High energy concentrate diets tend to result in lower rumen pH values, and pH buffers (e.g. limestone flour) are included in all the high energy feeds to alleviate this. Generally, high energy feeds are complete feeds (well balanced) and are designed for hard working cows, top producers, and to complement poorer quality roughages. Another common high energy feed is the super 25 concentrate, which gives 20% CP when mixed (i.e. in the final diet). It is formulated to accommodate the farmers' own source of raw materials. It is mixed in the ratio 70:30 with maize to equate to a super 20 feed (i.e. 70% super 25 mixed with 30% maize meal).

8 Silage making on-farm for dairy cows

Good quality silage is vital to maximise livestock production for both dairy and beef. In dairy production, good silage is normally equated with good milk yields and milk qualities. It is as simple as that. The better quality the farmer's silage is, the better the milk yield and components (milk quality in terms of fat content, total solids, solids not fat and protein content) will be. Silage is an invaluable addition to any successful dairy enterprise. It helps to maximise growth in young stock and gives optimal milk yield in all classes of cows.

The key to preserving silage at the highest quality is to ensure that the correct bugs (microbes) and enzymes go to work as quickly as possible to produce the acids needed to ferment the forage and prevent any further nutrient loss. Some commercial inoculants have been developed and farmers in the developing world are particularly encouraged to use these biotech products so that they can produce more from their animals.

Most of these inoculants are natural silage inoculants which provide these essential bugs and enzymes to ensure fermentation is reached in a minimum amount of time, with the least nutrients being lost in the process.

The high level of amylase enzyme included in some of the inoculants ensures that sugars are quickly made available for conversion to lactic acid. The following word equation illustrates the point.

- Maize – in presence of the enzyme, amylase → sugars available for conversion to lactic acid.

Two species of lactic acid bacteria (*L. plantarum* and *P. acidilactici*) have been selected for their ability to quickly and efficiently generate lactic acid from plant sugars. This ensures a rapid fall in the pH of the ensiled crop, meaning stability of ensiled feed is reached as quickly as possible and in-silo losses are minimised. This process can only occur if sugars are available in sufficient quantity.

- Plant sugars fermented by *L. plantarum /P. acidilactici* → lactic acid

So, the application of inoculants ensures more lactic acid and less of the undesirable products of fermentation are produced as illustrated in table 3.3.

Table 3.3 Effect of inoculant in lactic acid silage quality

	Untreated g/kg DM	Treated g/kg DM
Lactic acid (valuable end product)	31	36
Acetic acid (undesirable end product)	20	16
Alcohol (wasteful reaction)	52	45
Mannitol (unpalatable product)	41	29

New research shows that 50% more of the protein in forage is available when a biological additive is used. This explains why silage treated with an effective biological additive has better feed value than well fermented, but untreated, silage.

Maize silage provides today's dairy farmer with a high energy forage for use in most feeding situations. Likewise other grain silage such as sorghum, wheat and millet are becoming commonplace in certain regions. Mixed silages that incorporate grasses and legumes have been tried in parts of Southern Africa with promising results. These ensiled crops represent a major investment for any producer, and so maximizing fermentation once the crop is cut is essential.

Generally, inoculants as additives improve the speed and efficiency of both maize and whole crop silage fermentation. Controlling the fermentation is as important in maize silage as it is in grass silage. Nature will generally produce a reasonably well fermented maize crop, but the key to improving the efficiency of fermentation is by ensuring that plant sugars are converted to lactic acid (the desirable preservation acid), with the minimum of feed loss. Inoculants ensure that this lactic acid fermentation takes place, and that less desirable acids, ammonia and other substances that reduce palatability are kept to a minimum. Inoculants are aids to silage fermentation and the best results are obtained when good ensiling management is practiced. These good practices of silage management are:

- Correct, timely cutting of maize plants to be ensiled (i.e. at the milking stage)
- Adequate chopping of the maize plants or material to be ensiled
- Adequate compaction to exclude oxygen when filling the silage pit
- Correct siting and construction of the silage pit
- Protection from rain or seepage

Some inoculants are applicable for use on grasses, small grains and legumes, while others are applicable for use on maize and sorghum.

Farmers should note that silage making does not improve the quality of the roughage materials used, but that the process merely preserves the quality. The quality of the roughage used in the first place is vital. The "GIGO" effect applies (i.e. garbage in garbage out).

9 The problem of mastitis: its control using Bioplex zinc supplements

Mastitis is the inflammation of the udder system due to bacterial infection. The condition has been found to command significant financial losses in a dairy herd as shown in table 3.4.

Table 3.4 Breakdown of financial losses due to sub-clinical mastitis.

Type of loss	level of loss
(i) Death and premature culling	14 %
(ii) Discarded milk	8 %
(iii) Treatment	8 %

Udder health and Bioplex zinc

Trace minerals discussed earlier, such as zinc, copper, and manganese, play important roles in the immune system and tissue integrity. Zinc is needed in the formation of keratin, the wax-like protective lining in the teat canal. During each milking, up to 50 per cent of this first line of defence is washed out. If not totally replaced, the cow is vulnerable to infections such as mastitis. Keratin helps prevent micro-organisms (bacteria) that invade the udder through the teat end from causing these infections.

Several researchers recently showed how Bioplex zinc effectively reduced the incidence of mastitis in dairy cows. They found a significant decrease in new mammary infections when cows received Bioplex zinc as compared to cows fed the more commonly used inorganic zinc.

Somatic cell count

Researchers have also shown that feeding Bioplexes can reduce somatic cell counts (SCC). Danish commercial trials have shown significant reductions in herds with high SCC. Bioplexes have also been found to significantly reduce SCC in a herd with low initial SCC. Reproductive performance is improved due to reduced challenge on the cow's immune system as well.

Bioplexes are normally incorporated in feeds, liquid supplements, and mineral blocks.

10 Feed Additives available to dairy farmers

This section deals with feed additives available to dairy farmers. The various feed additives available to dairy farmers are yeast cultures (commercially known as *Yea-Sacc*), mycotoxin binders, bacterial additives, flavours, organic trace minerals, odour suppressants, and hay preservatives.

Yeast cultures

Diet specific yeast culture come commercially as *Yea-Sacc 1026* and *Yea-Sacc 8417* These are seeded into dairy feed, according to the manufacturer's specification which are normally not more than 10 grams per animal per day. The role in the diet is to:

- Buffer rumen pH,
- Increase cellulolytic and lactic acid utilizing bacteria numbers,
- Increase microbial protein flow,
- Increase feed intake,
- Increase milk production,
- Increase milk butter fat and protein and
- Improve persistency of lactation.

Mycotoxin binder

Mycotoxins are poisons produced by fungi that cause moulding of livestock feeds. Feeds, when damp or when the relative humidity is high, are susceptible to moulding. When animals feed on mouldy feeds, the toxins produced by moulds affect the animal's health. In dairy cattle, some of the toxins affect milk synthesis. As a result, it is necessary to find a way of binding these toxins because moulding that is mild is not obvious to the naked eye. Farmers feed their dairy cattle on mouldy material that contain mycotoxins without knowing. To protect their animals, certain substances that can bind the mycotoxins and renders them ineffective are widely used and these are generally known as mycotoxin binders. A good commercial mycotoxin binder is the *Mycosorb*. It is important to dairy feeding because it:

- Binds the poisons zearalenone and aflatoxin, and renders them ineffective,
- Improves intake,
- Improves milk production and
- Binds the bacteria salmonella and E. coli and attenuates them (bacteriostatic).

Bacterial supplements

Not all bacteria are harmful to livestock. Some bacteria are useful and are generally referred to as friendly bacteria. Bacterial additives which come in various commercial forms are useful to dairy cattle because:

- They help to stabilize the hindgut,
- They reduce winter scours/diarrhoea and
- They improve grain utilization.

Flavours

Flavours are important as additives to dairy feeds. This is because dairy cows respond to favourable smell and taste of the feed. The major advantages of flavour additives in dairy production are that they:

- Improve palatability of cationic/anionic salts, and
- Increase intake of the whole diet.

Organic trace minerals

Organic trace minerals are trace elements supplied, complexed, conjugated or bound to organic molecules, hence the name (see section 6). The most important and useful commercially available examples of organic trace minerals are Bioplexes, zinc-methionine complex and chromium yeast and selenium yeast. They are important in that they:

- Increase bio-availability of minerals,
- Increase bio-activity of minerals,

- Reduce somatic cell counts,
- Lower mastitis rates,
- Improve reproduction,
- Improve hoof strength, and
- Increase immunity

Odour suppressants

Odour suppressants are substances included in feed to suppress undesirable odours that affect feed intake. Odour suppressants help to improve or sustain high intake, milk production, reproduction and reduce ammonia odour in ammonia treated feeds.

Hay preservatives

Hay can easily go bad, especially when stored for a long time. Moulding is the biggest challenge farmers have to contend with. Preservatives that prolong the shelf life of the hay have been introduced on the livestock feed market and have helped farmers to preserve stocks of hay for future use. Commercial examples of hay preservatives contain buffered propionic acid. Preserved hay can be stored at between 18 and 20% moisture levels. The desirable characteristics of hay preservatives are that it should be non-corrosive, should reduce dry matter loss, reduce mould growth and dustiness. Some hay preservatives contain buffered propionic acid (and other acids) and have the following advantages:

- Reduce wastage of feed,
- Reduce heating,
- Inhibit mould growth and
- Their citrus flavour masks off odours.

11 Dry cow nutrition

A *dry cow* is one that has gone through a lactation cycle and is not producing any milk (i.e. literally dry). It will remain dry until it gives birth to the next calf, when another lactation cycle begins (i.e. dry period ends). Therefore, the dry period of a producing dairy cow extends from the end of one lactation to the start of the next, at the birth of a calf, and generally lasts 56 days (8 weeks). Nutrition over this period is important but often neglected and, as a result, both re-conception and milk production rates have been below expected levels, especially in the developing world.

The dry period is necessary for the cow to allow her body to regenerate the necessary secretory tissues (involution) in preparation for the next lactation, and also to replenish body reserves depleted during the preceding lactation. Involution of the udder tissue takes about 6 weeks, and if the cow does not complete this process, milk production may fall by as much as 30% in the subsequent lactation.

Remember, the dairy cow will be pregnant by now and it is also during this period that the growth and nutrition of the foetus is most rapid, and this places necessary emphasis on the nutrient provision over this stage.

Because of the high nutrient demand placed on the cow over the lactation period, body condition is often lost as requirements exceed supply from the feed. This is because intake levels are reduced during pregnancy and early lactation. This introduces the need for an adequate nutrient supply over the dry period to ensure the cow is physiologically able to cope with another lactation, and general veld grazing is not sufficient for this. Better quality forage and additional supplementation is often necessary.

Nutrition over the dry period is essential to ensure that at lactation, when there is a negative nutrient balance (i.e. the requirements are greater than can be obtained from the feed), the cow has sufficient body reserves to maintain production until intake increases and the nutrients can be sourced from the feed.

Conversely, a cow with too much condition (i.e. fat) will often suffer from ketosis as a result of being unable to break down the fat needed to meet energy requirements in the first few weeks of lactation. Optimum body condition should equate to a score of 3.5 (on a 5-point scale).

The dry period can effectively be divided into an early and a late stage, with differing nutritional requirements in each stage.

Feeding poorer quality forages to the freshly dried off cow is aimed at regenerating a healthy bacterial population within the rumen, which is often reduced as a result of the highly concentrated lactation rations required for maintaining high milk production. Protein levels at this stage should be in the region of 13%, which is not available from the grasses on the veld. Most commercial cow rations ensure that the necessary nutrients and protein is provided, while not opposing the need for a higher fibrous diet. During the close up period (nearing calving/lactation), the protein level is increased to 15%, and this allows the cow to begin adjusting to the lactation rations it will need for milk production, as well as ensuring the necessary nutrients are available for the foetus, during the third trimester.

Mineral requirements over the dry period must be considered carefully. In particular, it is important NOT to provide high calcium levels, as this will lead to a high incidence of milk fever at the onset of lactation. By providing the cow with calcium at this stage, she will have no need to initiate mobilisation of body reserves within the skeleton. As a result, at calving, when the calcium requirements are very high, the cow will be unable to utilize her calcium skeletal store soon enough, and the physiological effects of low blood calcium levels are seen (as milk fever).

This information calls for a closer look into the nutrition and feeding of the dry cow on the part of the farmer. Stock feed companies always strive to ensure that their *far off* and *close up* dry cow rations have the necessary levels of all the required nutrients that the animal needs over this period.

When and how to dry off the lactating dairy cow

Drying off should be done when the cow is 56–60 days away from calving down, or when milk production drops below economic levels. The process involves the removal of all concentrate feed, and a change of diet to a lower quality roughage. The low plane of nutrition ensures that the necessary levels of nutrients are inadequate for milk production, hence milk synthesis subsequently stops.

Drying off while cows are still producing around 20 kg of milk per day will result in the udder swelling. This in turn leads to dilated teats, which become vulnerable to infection through the exposed openings. Heavy udders can cause muscle damage, which may be permanent.

Cows are generally left for two milkings, and then milked at the third. This ensures that pressure in the udder is reduced, and that milk production has been effectively suppressed. Alternatively, an antibiotic dry cow remedy, such as *Shananast Dry Cow*, which is available from veterinary pharmacies, may be used. This remedy is inserted into each quarter after the final milking. The teat ends should be disinfected before administering the treatment and dipped afterwards. It is always advisable to check the antibiotic with your veterinarian or stockist for advice and approval.

13 Conclusion

Dairy production is a highly specialised area of livestock production and should always be treated as a science rather than an art. Apart from supplementing for nutrient deficits in growth, a dairy cow requires extra nutrients to synthesize milk and to reproduce another calf in order for the next lactation to come into being. The feeding of a dry cow is also a critical area that needs specialised attention, otherwise the future performance of the cow in terms of milk production and reproduction can be affected permanently.

References

Agriculture, Forestry and Fisheries Research Council Secretariat, *Japanese Feeding Standard for Dairy Cattle*, Chuouchikusankai, Tokyo, 1994

Bushnell, D.G, 'The importance and prevention of mould on groundnuts', in *Rhodesian Agricultural Journal*, 1964, **61**:108, pp. 11–34.

Department of Agriculture, *Handbook for farmers in South Africa*, Vol **3,** Pretoria.

Grant, R.J. and J.L Albright, 'Feeding behaviour and management factors during transition period in dairy cattle', in *Journal of Animal Science*, 1995, **73**:2791–2803.

National Research Council, Nutrient requirements of dairy cattle. Revised sixth edition, Washington D.C, National Academy of Science, 1988.

Roy, B, 'Calves: Give them dry feed as well as milk', in Farmers Weekly, 1988,October **21**: pp12–15.

Ruquin, H. and L Delaby, 'Effects of the energy balance of dairy cows on lactational responses to rumen – protected methionine', in *Journal of. Dairy Sciences*, 1997, **80**: 2513–2522.

Steve, M, 'Urgent need for mastitis control', in *Farmers Weekly,* 1988, October **21**: pp 16–19.

Sutton, J.D, 'Altering milk composition by feeding', in *Journal of. Dairy Sciences*. 1989, **77**:2801–2814.

Waltner, S.S., JP McNamara, and JK Hillers, 'Relationship of body condition score to production variables in high producing Holstein dairy cattle', in *Journal of. Dairy Sciences.* 1993, **76**:3410–3413.

4

SMALL RUMINANT FEEDS AND SUPPLEMENTATION

1 Introduction

Sheep and goat production in most tropical countries is based on grazing them on natural pastures and browse. In temperate environments, irrigated pastures are more important, especially with sheep. In both scenarios, forage quantity and quality become limiting factors in the dry season. Crude protein content of less than 7% is known to limit forage intake and animal performance. Seasonal variation in forage quality normally results in crude protein levels dropping to below 3% in the dry months and animals lose a lot of weight. This calls for the development of alternative protein sources (as supplements) of feed so as to reduce weight losses.

2 Feeding behaviour of goats

Goats feed on a wide variety of feeds. These range from tree and shrub leaves and grasses. Goats can easily differentiate between bitter, sweet, salty, and sour herbage. They are interesting in that they will refuse to eat feed that has been soiled by other animals. They also have a higher tolerance for bitter tasting feeds than cattle. It has been noted from experience that goats will not do well on a single type of feed for a protracted length of period. They display a high degree of selectivity and prefer a cafeteria type of feeding, such as a combination of grasses and shrub plants or tree leaves. Several goat specialists concur that palatability does not seem to be an important determinant of intake, but rather the availability of a variety of feeds. However, browsable material is the usually preferred feed type, because they can take a diet where 80% of the total intake is browse. Situational changes in preference easily occur where they are capable of feeding grasses and other crop residues like straws and stovers of cereals. Where goats have access to grasses, they prefer less coarse grasses, like guinea grass (*Panicum maximum*). They select against the coarser ones like elephant grass (*Pennisentum purpureum*). Grass pastures are better utilized if legumes are included. This boosts the nutritive value of the grass pastures. *Leucaena leucocephala* and *Cajanus cajan* have proved to be valuable feed for goats in the tropics.

In situations where grasses are provided to goats as a cut and carry diet, tree leaves should be provided as well. In most African environments, *Acacia* species and *Amaranthus spinosus* form important browse sources for goats.

3 Nutrient requirements

Adequate provision of clean water is a prerequisite for suckling goats and for those reared for mohair production. Those kept for meat require less water. Having said this, it is important to note that goats are capable of withstanding water shortages due to the fact that they have low water turnover rates and can resist dehydration. In comparison to sheep, goats under a hot environment of 36–38°C will pant at half the rate of sheep, sweat insignificantly and lose very little water in their urine and faeces. The animal can, therefore, go for days without consuming any water, during which urine excretion is severely reduced. Although goats show such remarkable ability to resist desiccation, it must be noted that the price of this physiological ability is a reduction in food intake. For maximum dry matter intake, a dry matter to total water intake of 1 to 4 should always be the target. Dry matter intake normally ranges from 2 to 3% of their live weight. The range is due to differences in production potential. An average diet of goats contains 8.4 MJ ME per kilogram of dry matter. With lactating does, the ME requirement should be about 10.5 MJ per kilogram of dry matter. The mean protein requirement for maintenance is 1.82 g digestible crude protein (DCP) per metabolic body work ($W^{0.75}$ kg), while for high producers it can be as high as 2.57 DCP per $W^{0.75}$ kg

4 Dairy goat rations

The basal diets are forages (foliage and grasses) and these may need to be supplemented with energy or protein concentrates to maximize production potential. However, concentrates should only be fed if they command more profits through, for instance, increased milk production. Concentrates, which contain 14% crude protein, are adequate when good quality grass pastures and good quality legumes are readily available.

The suggested ration per day for lactating goats weighing between 30–45 kg is shown in the following table:

Table 4.1 Suggested ration per day for lactating goats weighing between 30–45 kg

Scenario 1:	(i) Grass, or grass mixture (fresh)	2.5 kg
	(ii) Concentrates (17% CP)	0.6 kg
Scenario 2:	(i) Grass, or grass mixture (hay)	1 kg
	(ii) Concentrate (17% CP)	0.6 kg
Scenario 3:	(i) Grass with legume (fresh)	2 kg
	(ii) Concentrates (17% CP)	0.6 kg
Scenario 4:	(i) Grass with legume (hay)	0.8 kg
	(ii) Concentrates (17% CP)	0.6 kg

When feeding pregnant does, the amount of concentrate should be reduced to about 0.2–0.3 kg per day during the last week and then slightly increased to 0.4 kg per day, 2–3 weeks after kidding. Thereafter, concentrate can be increased or fed according to milk yield. Generally the concentrate should be provided at a rate of 0.4 kg per kg of milk yield.

A typical concentrate for goats giving 16–17% CP is composed of 18% cassava chips, 15% molasses; 65% groundnut cake, 1% iodised salt and 1% of mineral premix. For yearling and dry does, the feed, in the form of good pastures and legumes, should be *ad libitum* with 0.3 kg concentrate per doe per day. For breeding bucks, the concentrate level should be around 0.7 kg per day.

Kids need colostrum for 3 days right from day 1 and, at the latest, 4 hours after kidding. The colostrum can be provided from a nipple bottle, which is at body temperature. Weaning should take place at 3–4 months of age. By this time they should have been exposed to roughages and concentrates for at least 3–4 weeks so that they become able to nibble a sizeable amount per day for effective rumen development.

5 Feeding sheep

Sheep, like goats, consume a wide variety of feeds. Their productive efficiency is very low because half of the daily intake goes towards maintenance requirements. The low production efficiency is easily seen in fattening situations where the sheep have to be fed for an extended period to reach market weight, as compared to goats. A digestible crude protein content of 9% of the feed is adequate for body maintenance and 15% being adequate for maintaining and sustaining pregnancy, lactation and growth in lambs.

On average, daily intake on good quality grass and/or legume hay ranges from 0.9–1.5 kg for a live weight of 30–35 kg, supplemented with oil cakes as protein concentrates, 0,1–0,2 kg per day per animal.

A typical fattening ration for sheep is composed of 60% maize meal, 10% oil cake, 20 % brewers grain or silage, 4% cane molasses, 3% dicalcium phosphate, 2% bone meal, 0.5% salt and 0.5% vitamin/mineral premix. Adequate clean water supply should be guaranteed all the time. On average, sheep require 2 ml of water per gram of dry matter consumed. This means the water intake is in excess of 2 litres per animal per day.

6 Conclusion

In order to achieve excellent sheep and goat production and reproductive efficiency, farmers and animal scientists must focus on fortifying poor quality grazing and crop residues with molasses, urea and oil cakes. The high efficiency achieved is closely associated with cash flows that contribute to the profitability of herd reproductive programmes.

References

Devendra, C. and G.B Mclevoy, *Goat and sheep production in the tropics*, Singapore, Longman, 1992, p. 55–72 and 178–191.

Malachek, J.C. and F.D Provenza, 'Feeding behaviour and nutrition of goats on ranges', in *Proc. Int. Symp. On Nutrition and Systems of Goat feeding*, Tours, France, 12–15th May, 1981 pp. 411–428.

Mpofu, I.D.T., F Simoyi, and LR Ndlovu, 'Mineral status of goats under smallholder management', in *Transactions of the Zimbabwe scientific association*, 1998, **72**: 11–13.

Oyenuga, V.A. and AD Akinsoyinu, 'Nutrient requirements of sheep and goats of tropical breeds', in Proc. 1st Int. Symp. Feed composition, Animal Nutrient Requirements and Computerisation of Diets, Utah Logan, 1976, p505–511.

Sibanda, L.M,. 'Small ruminant production in Zimbabwe', in Prospects and constraints. Proc., of workshop at Matopos Res. Station, Zimbabwe Small Ruminants Network, 19–20 August, 1993

Ward, H.K, 'Some observations on the indigenous ewe', in Rhodesia Agriculture Journal, 1959, 56:218–223.

5

POULTRY FEEDS

1 Introduction

Feed represents the largest single expense in poultry production and accounts for approximately 80% of the total cost of production. It is, therefore, imperative that poultry rations be nutritionally balanced and properly mixed for the economical production of poultry meat and eggs. Feed must provide the necessary proteins, carbohydrates, fats, minerals and vitamins both in quantity and quality, as cheaply as is possible.

With this in mind, the stock feeds industry has a range of poultry mineral/vitamin premixes, concentrates and complete feeds for both broiler and layer chickens using least cost ration formulation procedures. The concentrates with mineral/vitamin premixes allow for partial mixing of poultry feed on farms with locally sourced maize, or sorghum and soya, thus enabling producers to save input costs by using on-farm feed ingredients. Complementing these feeds is a range of poultry feed additives, which include carotenoids, stress packs, eggshell strengtheners, growth enhancers and anticoccidials.

2 Feeding layers

Chicks

From day old to 8 weeks of age the chicks must use commercial grade chick mash or chick broiler concentrate. The farmer should mix 2 parts concentrate to 3 parts maize by weight. Both feeds normally contain a coccidiostat. To calculate feed requirements, farmers are encouraged to work on approximately 1.8 to 2 kg feed per bird (i.e. up to 8 weeks of age).

Growers

From 8 to 18 weeks, the birds must be fed growers mash or growers concentrate, by mixing 2 parts concentrate to 3 parts maize by weight. Again the feeds normally contain a coccidiostat. To calculate requirements, producers are supposed to work on approximately 4.5 kg of feed per bird (i.e. for the period 8–18 weeks).

Layers

From point of lay, i.e. 18 weeks of age onwards, medium energy layers mash or pellets should be fed to heavy hybrids, pure-breds or backyard poultry. High energy layers mash or pellets is fed to all other layers. Birds should consume approximately 125g of feed per day or 45 kg per year. Layers concentrate is made by mixing 2 parts concentrate to 3 parts maize by weight and can be fed to all types of layers. All mixed poultry feeds are completely balanced in the total nutritional requirements of poultry – if procured from a genuine feed manufacturer.

3 Management hints for layers

Feeding

Layers should normally be fed ad-lib (i.e. without any restrictions). Feed should be available at all times. However, heavier breeds may require rationing to prevent over-consumption that may result in birds that are over-fat. One tubular feeder should be supplied per 25 birds or 10 to 15 cm feeding space per bird. Any sudden change in feed quality adversely affects egg production. The feed should be gradually (making sure that each day the change in proportion of feed does not exceed 15%) introduced over 7 to 10 days. As the cost of feed accounts for up to 80% of the total cost of production, feed wastage should be avoided at all cost.

Drinkers for poultry

Cool, fresh water must be available at all times. The drinkers should be washed out at least twice a day and refilled with fresh water. Spillage should be avoided to avoid damp litter. If the drinkers are movable they should be placed in a new position each day. Layers require 5 cm of drinking space per bird, and the depth of the water in the drinker should be 2.5 cm. 100 layers will consume approximately 25 litres of water per day. Inadequate drinking space will severely depress egg production.

Litter management

Use only fresh, dry, absorbent material, e.g. chopped hay/grass or wood shavings. Initially, the absorbent materials must be placed to a depth of 10 cm and more could be added as and when required to give a depth of 20 to 25 cm at the end of the day. It must be ensured that the litter is dry and friable at all times. Caking must be avoided and all wet litter must be removed without delay.

Floor space management

The area required depends on the size of the bird. Light breeds require 0.23 to 0.28 metres square and heavier breeds 0.33 to 0.37 metres square. Over-crowding will give rise to cannibalism, damp litter, poor health, higher mortality, depressed feed intake and hence depressed egg production.

Nests

It is recommended that the farmers provide the nests measuring 30 cm x 30 cm x 30 cm, per 4 to 6 hens. Nesting material of chopped hay/grass or wood shavings must be kept clean at all times to avoid dirty eggs. Eggs should be collected every two hours to prevent egg eating, dirty eggs and broodiness. Broodiness is a sex linked characteristic exhibited by the layers getting into the mode of incubating their eggs. This results in a large drop in the number of eggs produced by the flock. Any bird showing signs of broodiness should be isolated into an unfamiliar environment for five days until the behaviour goes down. The hen will normally resume laying after 25 days. Gonadotrophic hormones can be injected and these reduce broodiness. However, stockmanship methods are more effective.

Perches

Perches should be placed or mounted approximately 60 cm above the ground, allowing 30 cm of perching space per bird. The perches should be 35 to 40 cm apart. It must be ensured that the woodwork is well disinfected and creosoted to avoid external parasites.

Lighting

To maintain satisfactory production, it is essential to supply artificial lighting, particularly during the winter months (hence the day length increases artificially). Increasing the day length from 14 hours at onset of lay to a maximum of 16 to 17 hours must be done gradually over a period of 8 to 10 weeks.

General

Attention to detail is essential at all times. Birds must be protected and records kept up to date always. All mortalities must be examined to ascertain cause of death. Remember prevention is better and cheaper than cure. Maintaining a high standard of hygiene and cleanliness is the answer. Once birds have been removed from a house, it is essential to wash the pens with clean water, disinfect, whitewash all brickwork, creosote all woodwork and leave vacant for a minimum of 14 days in order to break the life cycle of diseases.

4 Feeding of broilers

Broiler starter

During the first four weeks, the producers can feed 1 kg per bird of super broiler starter or super broiler concentrate. The farmer is advised to mix 2 parts concentrate to 3 parts maize by weight. Once the food has been consumed, a change should be made to super broiler finisher or super broiler concentrate by mixing 1 part concentrate to 2 parts maize by weight. To calculate the requirements of super broiler finisher, farmers should work on approximately 3.3 kg of feed per bird during 5–7 weeks of age.

OR

From day old to 4 weeks of age, broiler starter or chick/broiler concentrate is fed by mixing 2 parts concentrate to 3 parts farm produced maize by weight. To calculate the requirements, work on approximately 1.3 kg of feed per bird as a standard guide. From 4 weeks of age to slaughter, broiler finisher or chick/broiler concentrate is then fed by mixing 1 part concentrate to 2 parts farm produced maize by weight. To calculate the requirements, working on approximately 3.5 kg of feed per bird up to 8 weeks of age is recommended.

All broiler commercial feeds normally contain a coccidiostat. All mixed poultry feeds are completely balanced in the total nutritional requirements of poultry.

Management hints for broilers

Heat management

It is necessary to keep the young birds warm at all times. Over heating in the first couple of weeks is preferable to chilling. Brooder temperatures are shown in the table below:

Table 5.1 Recommended brooder temperatures are:

First week	32 to 35 degrees Celsius
Second week	27 to 32 degrees Celsius
Third week	24 to 27 degrees Celsius
Fourth week	21 to 24 degree Celsius

Careful observation of the birds' behaviour will indicate whether the brooding temperature is correct. If they crowd in a corner they are probably chilled, and if they space out too much and are restless, they are feeling too hot.

Feeding broilers

Broilers must have feed present in the feeding troughs at all times. The feed should never be rationed. For the first few days, it is advisable to place the feed on egg trays or shallow containers. Thereafter, depending on age, broilers will require from 5cm to 10 cm feeding space per bird or 1 tubular feeder per 25 to 30 birds.

As the cost of feed accounts for up to 80% of the total cost of production, wastage should be minimized or, better still, prevented.

Drinkers

As with the layers, cool, fresh water must be available at all times. The drinkers should be washed out at least twice a day and refilled with fresh water. Every effort should be made to avoid spillage as this results in damp litter. If the drinkers are movable, they can be placed in a new position each day. Depending on age, broilers will require 1.5 to 2.5 cm drinking space per bird. As a general recommendation, drinkers should not be more than 3m away from the feeder.

Floor space

The recommended floor area is 0.093metres squared per bird or 9.3metres squared per 100 birds. Over crowding will give rise to cannibalism, increased mortality, poor health, depressed feed intake and, therefore, poor growth rates.

Litter

In deep litter systems, only fresh, dry, absorbent materials, e.g. chopped hay/grass and wood shavings should be used. They should be placed to a depth of approximately 10 cm and fresh litter added when required. The litter should be dry and friable at all times. Caking should be avoided and all wet litter must be removed.

General

Attention to detail is essential. Birds should be observed frequently. All mortalities must be examined to ascertain cause of death. Remember prevention is better and cheaper than cure. Maintaining a high standard of hygiene and cleanliness is a virtue. It is next to godliness. Once the birds have been removed from a house, the house should be washed down with water, disinfected, brickwork whitewashed and the woodwork creosoted. They then should be left vacant for a minimum of 14 days to break the life cycle of disease-causing micro-organisms.

5 Home mixing for poultry enterprises

Due to the escalating costs of poultry production, an increasing number of producers are shifting towards partial or complete mixing of their own feed on the farm. Bought in concentrates, maxipacks and other ingredients are mixed with farm-grown or locally sourced cereals, usually maize, sorghum or millet; and sometimes with cereal by-products such as maize bran or maize germ. Under these circumstances, quality control is very important. Home mixers should check the quality of their cereal grains for mould and overall physical appearance.

Cereal by-products have a shorter shelf-life compared to cereal grains, as most of them tend to go rancid if they are kept for too long before use. The presence of 'off' smells in these by-products is a good indicator of a product which has gone rancid. Rancidity affects growth and performance in poultry. Furthermore, expert advice should be sought on how best to combine these maize substitutes when home mixing with bought in concentrates.

6 Typical poultry rations

Table 5.2 presents typical poultry rations for growers and layers which can be adopted by the 'do it yourself' farmer, as long as raw materials are available, and that weighing and mixing respectively are done accurately and properly. Farm gate preparations for day old chicks are strictly discouraged. Commercial grade chick mash or concentrate should be used always from day old up to 3–4 weeks of age.

Table 5.2 Poultry ingredient levels per tonne of feed

Ingredients	*Growers kg/tonne*	*Layers*
Maize meal	600–650	600–650
Maize meal chaff or wheat feed	100–105	150–155
Fish meal or meat and bone meal	100–120	80–100
Soya bean meal	80–85	80–85
Limestone	10	30
Phosphorus source (dicalcium phosphate)	5	5
Salt	2	2
Vitamin & mineral premix	3	3
Other ingredients (i) coccidiostats (ii) anti oxidants (iii) antibiotics		
Carotenoid pigment	No	Yes
Total	1000	1000

The composition of mineral premix for one tonne of broiler or layer diet should be maintained according to table 5.3.

Table 5.3 The composition of mineral premixes

Element	*Amounts Per Tonne*
Manganese sulphate	332.0 g
Iron sulphate	205.0 g
Zinc carbonate	86.0 g
Copper sulphate	10.0 g
Potassium iodide	1.5 g
Cobalt sulphate	0.5 g
Sodium selenate	0.5 g
Inert filler, difference to make a total of 1 kg	[ ]
Total	*1 kg*

The composition of vitamin premix for one tonne of broiler or layer diet are given in table 5.4 .

Table 5.4 Composition of vitamin premix

Vitamin	*Amounts per tonne*
Vitamin A	6 400 000 IU*
Vitamin D3	1 600 000 IU
Vitamin E	6 400 mg
Thiamin	1 500 mg
Riboflavin	3 200 mg
Niacin	16 000 mg
Vitamin K	240 mg
Pantothenic acid	10 000 mg
Folic acid	500 mg
Vitamin B12	8 mg
Choline chloride	1 000 g
Filler	[ ]
Total	*2.0 kg*

Mineral/vitamin packs usually come pre-packed (hence the word premix) and are available from stock feed and pharmaceutical companies. Sometimes, in addition to mineral/vitamin premixes, these packs may also contain limestone flour, monocalcium phosphate and additional amino acids to make a bigger pack (30–50 kg) called a maxipack. These maxipacks are mixed with soya cake and maize meal to make complete feeds.

* International Units:
One IU of vitamin D = 0.025 μg crystalline vitamin D
One IU of vitamin A = 0.3 μg of vitamin A

7 Conclusion

Broilers and layers are usually fed to appetite. Breeding stock is an exception to this general rule. Otherwise feed requirements are expressed as the proportion of particular nutrients a poultry diet must contain. Feeding breeding stock without any restrictions results in over-fat and hence infertile birds.

References

Anthony, P, 'Programme for prolific pullet', in *Farmers Weekly,*. Durban; October 21, 1988, pp 10–11.

Dean, W.F. and Scott, 'The development of an amino acid reference diet for the early growth of chicks', in *Poultry. Science,* 1965, **44**:803–805.

Emmert, J.L. and D.H Baker, 'Use of the meal protein concept for precision formulation of amino acid levels in broiler diets', in *Journal of. Applied. Poultry*, 1997, **6**: 462–464.

Han In, K. and J.H Lee, 'The role of synthetic amino acids in monogatsric animal production', in *Asian-Australian. Journal of. Animal. Science*, 2000, **13(4)**: 543–560.

Kornegay, E.T., ed, 'Advances in amino acid nutrition and metabolism of swine and poultry', Baker, D.H. in *Nutrient Management of food Animals to Enhance and Protect the Environment,* New York, CRC Press, 1996, 41.

Lipstein, B. and S,Bornstein, 'The replacement of some of the soya bean meal by the first limiting amino acids in practical broiler diets. 2. special addition of methionine and lysine as practical substitutes for protein in finisher diet', *British. Poultry. Science,* 1975, **16**:189–190.

Mutetwa, L, 'Production of Quality Poultry Feed – Some Salient Points', I.N.Z. Bulletin, September 2001.

NRC, *Nutrient requirements of poultry*, ninth edition, Washington D.C National Academy Press, 1994

Topps, J.H. and J Oliver, 'Animal Foods of Central Africa', in *Technical Handbook No. 2, Zimbabwe Agricultural Journal, 1993,* p 32.

Uzu, G, 'Thresomine requirements in broilers', in *Alimentation equilibre commentary, Document No. 242.* Commentry, 1986

Waldroup, P.W, RJ Mitchell, JR Payne, and KR Hazen, 'Performance of chicks fed diets formulated to minimize excess levels of essential amino acids', in *Poultry. Science.* 1976, **55**:243.

6

PIG FEEDS

1 Introduction

To obtain the best possible production from a pig unit, a high level of management combined with the optimal nutritional regime is required. A wide range of pig feeds are available to provide this optimal nutritional regime that different classes of pigs require.

2 Complete pig feeds

Brood sow meal

Brood sow meal is a complete ration containing about 15% crude protein and all the essential amino acids, vitamins, minerals and trace elements that a breeding stock requires in order to perform to its maximum potential. Breeding stock should be fed according to their body condition. Boars in good condition should be fed approximately 1.8 kg to 2.0 kg of brood sow meal per day. Sows and gilts will perform well on 2.0 kg per day. However, it may be beneficial to increase this in cold weather.

Pig creep pellets

Piglets grow on both mother's milk and supplementary feed (creep feed). Pig creep Pellets (22% CP) are formulated to maximise the intake of piglet feed. By so doing, a more economical porker or baconer is produced. Pig creep pellets should be fed *ad lib* from day 7 to day 56, after which piglets are moved onto pig growth meal. This change over should be done gradually over a 7 day period. A smooth change over will ensure diarrhoea-free operations in the grower pens.

Pig growth meal

Pig growth meal (18% CP) is formulated to allow the best possible feed conversion from pigs, thereby improving the profitability of pig operations. This is a well balanced diet to be fed to growing pigs from 8 weeks of age until they attain 60 kg live weight. At 10 weeks of age, it is recommended to restrict the intake to 1.3 kg per pig per day. From week 12, the feed should be increased weekly by 110 grams per pig until a maximum feeding level of 2.0 kg per pig per day is reached in the summer months and 2.3 kg per pig per day in the winter months.

Pig finisher meal

This is a complete feed that is normally fed to pigs of 60 kg live weight. Pig finisher meal (16% CP) promotes lean tissue growth, allowing for better market grades. Pigs should be fed 2.0 kg to 2.3 kg per day.

3 Pig concentrates

Pig creep concentrate

Pig creep concentrate is designed to be mixed with milled maize at a ratio of 2 parts pig creep concentrate to 3 parts milled maize by mass. The resultant mix gives a balanced feed containing at least 19.0% crude protein which should be fed *ad lib* to young piglets from 7 days to 8 weeks of age.

Pig growers/finisher concentrate

Pig grower/finisher concentrate is designed for the farmer who has his own maize. Pig grower/pig finisher concentrate when mixed at a ratio of 2:3 (i.e. 2 parts concentrate to 3 maize meal) with milled maize by mass, will result in a ration containing all the necessary vitamins and minerals required by a growing pig.

The resultant mix will allow pigs to produce at their maximum efficiency.

Dry sow concentrate

Many commercial feed manufacturers have designed a concentrate for the modern sow to use with snap corn or milled maize at a ratio of 2:1 by mass. This maintenance feed is recommended for high feeding allowance between weaning and re-mating, maximising the change of a prompt return to oestrus, strong heat and high ovulation rate. Sows should eat 3–4 kg per day with intake gradually falling as the sow approaches oestrus. Using a bulky dry sow diet encourages a higher gut fill during pregnancy. Overfeeding is not recommended during this period.

Brood sow concentrate

This concentrate is designed for use with snap corn or milled maize at a mixing ratio of 2 parts brood sow concentrate to 3 parts milled maize by mass. The mineral and vitamins needed by the lactating sows are included in the concentrate in appropriate ratios. This feed, which is ideal for gilts and boars, improves the sow efficiency if fed liberally at strategic points in the cycle.

Feeding recommendations for the brood sow concentrate are shown in table 6.1:

Table 6.1 Feeding recommendations for the brood sow concentrate

Pregnant sow	2 kg per sow per day
Dry sow (weaning to service)	4–6 kg per sow per day
Gilt	2.7 kg per gut per day
Lactating sow	2 kg per sow + 0.5 kg per piglet being suckled

Lactating sow concentrate

A concentrate mixed with milled maize at a ratio of 1:1 by mass will give a lactating diet that will ensure adequate milk supply for the piglets. As a rule of thumb, each additional kg of lactating food consumed per sow per day increases weaning weight by 0.55 kg per piglet. The nutrient requirements for the lactating sow are vastly different from dry sows. Therefore, it is recommended that two different feeds are used during the two cycles the sow undergoes.

4 Urea warning in pig feeding

All urea-containing feeds are potentially dangerous for pigs. The normally recommended precautions against urea poisoning should be taken and antidotes should be available for rapid administration if necessary. High value breeding stock should not be fed urea-containing feeds, as losses to the pig unit can be catastrophic.

Raw materials for stock feeds are carefully selected for their nutritional value, palatability and cost effectiveness. The raw materials and finished feeds should be regularly tested in commercial laboratories to check for nutrient specifications and presence of anti-nutritional factors.

5 New products

Based on new research findings and to meet changing market requirements, new products are constantly being developed, tested and added to the feed ranges by feed manufacturers. Feeds can also be customer formulated to suit specialised requirements.

6 General guide to pig feeding

Management hints

The most important factor affecting the profitability of a pig enterprise is the number of piglets reared per sow per year.

Producers should aim at a minimum of 20 piglets per sow per year. The piglet has a body temperature of 39–40°C and hence requires a comfort zone of 34–40°C in the creep area. The creep area temperature should be decreased approximately at a rate of 1°C per week until 8 weeks of age when the temperature should be about 23 to 24°C.

As a general guide, the following target weights can be used to see if temperature conditions are optimum:

- Birth weight should be 1.4 to 1.5 kg
- At 3 weeks old the piglets should be 5.5 to 6.0 kg live mass
- At 5 weeks old the piglets should be 7.5 to 9.0 kg live mass
- At 8 weeks old the weaners should be 18 to 20 kg live mass

It is imperative to ensure that all piglets get colostrum within 6 hours of birth. At birth the navel and umbilical cord should be treated with iodine and the hooves dipped in iodine to prevent bacteria from entering the soft tissue. Teeth are clipped carefully, leaving enough teeth length. Clipping too low will crack the tooth. In some countries teeth clipping is considered inhuman as a stressor, so the clipping of teeth is discouraged. Administration of iron injection in the side of the neck, rather than the hind leg, should be the standard practice with piglets on day 2. This is done in order to prevent anaemia.

The herd health programme should involve the following, because feeding healthy animals is key to profitability:

- De-worming and treating the sow for mange at about 7 days before entering the farrowing house. After weaning, udders are washed and teats clipped to prevent udder infection. The sow is also given a *parvo* vaccination 20 days before service and 14 days after furrowing,
- The gilt should receive both *parvo* and erysipelas vaccinations at about 85 kg body mass. These two vaccinations are to be repeated three weeks later, and
- The boar should receive erysipelas vaccinations twice a year after the initial vaccinations. The animals should also be treated regularly for mange.

How to estimate feed requirements for pigs

Actual feed consumption figures for pigs are extremely variable and depend on numerous factors: e.g. age of sow; number of piglets per litter; whether restricted or *ad lib* feeding; age of weaning; type of housing, production of porkers, baconers or heavy hogs; group sizes and sexes and breed.

However, as a very basic guide, the following information listed in table 6.2, may be of assistance to pig producers.

Table 6.2 Basic guide for feed budgeting in pig production

Type of pig	*Feed intake /day/animal*	*Duration/days*	*Total feed required*
Pregnant sow	3 kg	114 days	342 kg
Lactating sows with 10 piglets	1.5 kg + 0.5 kg/piglet)	42 days	273 kg
weaning to service	6.5 kg	21 days	136.5 kg
Grower/ finisher	ad lib	up to 90 kg live mass	260 kg
Piglets/creep stage	ad lib	<42 days	20 kg
Boar	2.7 kg	–	–

As a general guide, creep feed constitutes 5.5% of the total feed costs, sow and boar meal 21% and growers and finishers, 73.5%. Considerable saving in feed could be achieved by weaning at an earlier age i.e. 3–5 weeks, and by getting the sow into pig sooner, i.e. 14 days. In addition, this will increase the ultimate number of farrowings per year.

7 Conclusion

The basic aim in pig production is to encourage the fastest possible growth rate without excessive deposition of fat. This may be achieved by feeding them to appetite, especially for Large White and Landrace breeds.

References

ARC, 'The nutrient requirements of pigs', in Agriculture Research Council. Commonwealth Agriculture Bureaux, Slough, UK, 1981

NRC, Nutrient requirements of swine, tenth edition, Washington D.C, National Academy Press, 1998.

Pettigrew, J.E, 'Amino acid nutrition of gestation and lactating sows', in Biokyowa Technical Revie, No. 5, 1993, pp. 1–8.

Stahly, T.S, 'Impact of immune system on growth and regimens of pigs', in: Recent Advances in Animal Nutrition, 1996, pp. 197–206.

SCA, 'Feeding standards for Australian Livestock (Pig)', Standing Committee on Agriculture, Pig subcommittee, 1987.

Tokach, M.D. RD Goodband, and JL Nelssen, 'Recent developments in nutrition for the early – weaned pig', in Common Contin. Educ. Pract. Vet, 1994, **16**: 407–409.

Wang, T.C. and MF Fuller, 'The optimum dietary amino acid pattern for growing pigs', Br. J. Nutr, 1989, **62**: 771–779.

Williams, N.H. and TS Stahly, 'Impact of immune system activation on the lysine and sulphur amino acid needs of pigs', in ISU Swine Research Report, Ames IOWA State University, 1995, pp. 31–34.

7

OSTRICH FEEDS

1 Introduction

Stock feeds companies manufacture a range of ostrich feeds to suit specific needs. Their feeds are available in meal or nut form. Except for the complete starter and complete maintenance rations, roughage needs to be added to the other ostrich feeds. Producers can, therefore, adjust the concentrate to roughage ratio, depending on the bird condition. Ostrich concentrates are well suited to Rhodes grass, although a wide range of roughages are compatible. This makes economic sense when the variety of production systems is taken into account, and a farmer can feed according to particular conditions, i.e. intensive, semi-intensive or extensive.

2 Profitability of ostrich farming

Live chicks are what counts in a profitable ostrich enterprise. Good nutrition and hygiene are vital to ensure their survival. Feed should be available on an *ad-lib* basis and clean water should be supplied at all times. Good chick management will ensure a strong foundation for future profit. Minimising feed cost should be the guiding principle in order to guarantee profitability.

3 Ostrich concentrates

Complete starter crumble

Chicks should be fed the complete starter crumble from the first day to 2–3 months of age. No mixing is required as this is a complete feed with the right amounts of fibre included. The farmers may add small amounts of finely chopped green fibre to attract chicks initially. Ostrich feed on colour rather than taste. Cool clean water supply, which is accessible to the chicks, should be available all the time. Feeding of the complete starter crumble is *ad-lib*. Change over to growers meal should take place gradually over a period of 2–4 weeks.

Grower meal/nuts

Grower meal or nuts suits the ostriches from 2–3 months to 6–8 months of age. The meals or nuts should be mixed in the ratio of:

- 3 parts grower meal to 1 part dry fibre by weight, or
- 3 parts grower meal to 5 parts green fibre.

Feeding should be *ad-lib* in order to attain maximum intake and accelerated growth. As birds mature, they are able to efficiently digest increasing amounts of fibre in their diets.

Breeder meal/nuts

Breeder meal is fed to mature stock a month before egg laying is due to commence and throughout the laying period. The meal is fed with at least 10% good quality roughage on a dry matter basis (green if possible). Normally, feed is given at the rate of 1.5–2.5 kg of breeder meal/nuts per bird per day, depending on production system and condition of the birds.

Maintenance meal/nuts

Maintenance meals, which are roughage free, are normally fed to birds during the non-breeding season. Mixing is done at the ratios of:

- 9 parts maintenance meal to 1 part good quality dry fibre by weight.
- This equates to a 14% CP complete feed.

At least 1–2 kg per bird per day is recommended for most indoor feeding systems.

4 Range of ostrich complete feeds

Ostrich starter mash/crumbs

The usual chemical composition of ostrich starter mash is approximately CP 20.0%; CF 6.6%; Calcium 1.4% and phosphorus 0.8% on 'as fed' basis. Ostrich starter mash should be fed *ad-lib* to chicks up to eight to twelve weeks after hatching. The addition of not more than 5% of finely chopped green roughage as an attractant is optional. Moistening and stirring the feed by hand stimulates interest in consumption. Grit or fine gravel should be made available in the play area in small amounts. When the chicks pick them up, the stones will act as an aid to digestion.

Ostrich grower mash/cubes

The growers mash is composed of:

- Crude protein (CP) level of 15%
- Crude fibre (CF) level of 8.4%
- Calcium level of 1.1% and phosphorus level of 0.8%

Ostrich grower mash/cubes should be introduced as part of the chick's ration in a gradual change from ostrich starter at about eight weeks of age. By twelve weeks of age the chicks should be at their target weight and feeding *ad-lib* on the ostrich grower mash/cubes. Gravel or small stones should be available in the pens or play area and good quality roughage provided for older chicks and sub-adults. Ostrich grower mash/cubes may be used, as a maintenance ration, for non-breeding adult ostriches, at the rate of 1–1.5 kg per bird per day.

Ostrich maintenance mash/cubes

The maintenance mash has the following chemical composition:

- CP 12%
- CF 14%
- calcium 1.2% and phosphorus 0.7%.

Ostrich maintenance mash/cubes is fed to sub-adult and non-breeding adult ostriches at a rate of 1–1.5 kg per bird per day. Access to gravel or small stones should be provided.

Ostrich breeder mash/cubes

The breeder mash is composed of

- CP 15%
- CF 9.4%
- Calcium 2% and Phosphorus 0.7%.

Ostrich breeder mash/cubes is fed to breeding adult ostriches at a rate of 1.5 to 2 kg per bird per day. Good quality roughage should be provided and access to gravel or small stones should be allowed.

5 Feeding management

If a situation arises that makes diet switching inevitable, the change from the current feed to another should be done gradually over a 10 day period. The farmer should slowly increase the proportion of new feed until 100% substitution is attained by day 10. The farmer should ensure that the new feed should be increased by no more than 15% at a time. The ostrich producers should always bear in mind that ostriches are prone to proventriculus impactions. Impactions occur when normal peristalsis of digesta is affected and the ingested feed sediments in a section of the alimentary canal, making the animal feel full (pseudo-satiety). It normally occurs when the birds have been stressed in one way or the other. Reducing stress and encouraging exercise will decrease the incidence of impactions.

Impactions can be treated if identified early enough by flushing the alimentary canal with tap water while holding the bird upside down. This operation must be done by experienced farm operatives or with the help of veterinary personnel. If the farmer constantly moves birds from one location to another, either within the same production unit or to a different production unit, the bids will be stressed. The farmer must also carefully check the pens in which the birds will be placed. Any object that can be picked up and swallowed should be removed. Feed should not be changed at the time the birds are moved. Feed should be changed either 1 or 2 weeks before or 1 to 2 weeks after the birds have been moved.

Exercise is very important with ostrich birds. Exercise plays an important role in preventing leg problems, especially in young birds. Birds should have as much space as possible in which to walk and run. The earlier the birds are allowed to move and exercise, the lower the probability of leg problems in their lifetime.

Raw materials for ostrich feeds should be carefully selected for their nutritional value, palatability and cost-effectiveness. The raw materials and finished feeds should also be tested regularly where necessary and possible.

Based on new research findings and to meet changing market requirements, new products are constantly being developed, tested and added to the range of ostrich feeds.

6 Conclusion

Ostriches need careful feeding. A high plane of nutrition produces many fertile eggs. However, too much feed leads to over-fat birds and this results in infertile eggs. Any feed ration should provide minimum requirements for critical nutrients.

References

Billett, 'Ostrich feed lotting', in *Farmer's Weekly*, South Africa, July 13, 1984, pp. 28–32.

Hallan, M.G, 'The Topaz introduction to practical ostrich farming', Harare, 1992, pp. 53–60.

Hastings, M.Y, 'Ostrich farming', Armidale, University of New England Printery, 1991, pp. 33–40.

Smith W.A. and J Sales, 'Practical guide for ostrich management and ostrich products', Alltech, Stellenbosh University Printers, 1995, pp. 8–19.

Swart, D., RI Mackie, and JP Hayes, 'Fermentative digestion in the ostrich (Struthio camelus var. domesticus), a large avian species that utilizes cellulose', in South Africa Journal of Animal Science, 1993, **23**: 119.

Van Niekerk, B, 'The Science and Practice of Ostrich nutrition', AFMA FORUM, Sun City, 7–9 June 1995.

Vohra, P, 'Information on ostrich nutritional needs still limited', in *Feedstuffs,* 1992, **64**:16.

8

HORSE FEEDS

1 Introduction

A lot of mystery surrounds the feeding of horses. The horse is not a ruminant, though it is often fed as if it is. Neither is it a monogastric despite its physiology. Horses, in fact, are semi-ruminants. It has become evident in horse nutrition that the complete range of vitamins (A, B_{12} C, D, E and K) and macro-elements (Ca, P, Mg, Cr, Na, K, S and I) must be present in a form which is biologically available and sympathetic to the digestive system of horses.

2 Feeding guide

A correctly balanced and consistent diet is essential for the well-being of horses. Feed is digested more efficiently if given in small portions at regular intervals rather than once or twice a day, while high quality hay or grass should make up half to two thirds of the total daily diet. Daily allocation should be determined by live mass, workload and, in the case of breeding mares, stage before foaling down.

Race horse meal or nuts are a high quality, high performance diet, formulated to meet the demands of the athletic horse, in terms of energy provision and endurance to fatigue.

This diet contains chromium, which is an essential mineral recommended as a means to increase stress resistance and stamina, and to overcome lactate build-up during strenuous exercise or high speed racing. Feeding of racehorse meal is recommended at 1–2% of body weight. Good quality hay should be freely available all the time.

Horse conditioner meal or nuts are an all purpose diet, highly recommended for improving and maintaining the condition of all types of mature horses, including broodmares and lactating mares. Feeding is recommended at 0.5% to 2% of body weight. Good quality hay should be freely available all the time as well.

Horse concentrate, when mixed with maize on the farm, equates to the horse conditioner. Mixing is done in the ratio of 65% concentrate to 35% maize by mass. Feeding is recommended at 0.5% to 2% of body weight. Good quality hay should be freely available.

Sweet horse (11% CP) mix is a low cost, high molasses, general purpose horse feed designed to meet maintenance requirements during the off season and for non-working horses. Feeding is recommended at 0.5–2 % of body weight. Good quality hay should be freely available.

High protein sweet horse mix

High protein sweet horse mix at 16% CP, is a high protein variation of sweet horse mix formulated to maintain fitness in the sport horse, e.g. polo and polocrosses. It is recommended that the addition of maize should be done to this feed to give extra energy when horses are working. Feeding an extra 0.5 kg of maize when the horse is in light work and an extra 2 kg of maize when the horse is in heavy work produces the desired condition and performance on the track.

Feeding lush protein sweet horse mix (16% CP) at 1–2% of body weight and ensuring that good quality hay is freely available is the rule of thumb.

3 Feed additives for horses

Stock feed companies produce a range of high quality premixes and feed additives for horses. They supply a standard premix and a high performance premix for home mixing operations. Some of the feed additives available include horse electrolyte, vitamin E, selenium horse supplement, equivite, and nutriquine.

Horse electrolyte

When a horse sweats, essential body salts (electrolytes) and fluids are lost and this may cause fatigue and muscle stiffness, thereby prolonging recovery time. Horse electrolyte is a mineral and salt replacer that should be fed to horses in training and at full work. For horses on light work, 50 g of electrolyte should be fed every second day. This should be mixed in feed or 4 litres of drinking water. For heavy work, horses should receive 50 g of electrolyte every day and this is mixed with feed or with 2 litres of drinking water.

Vitamin E and selenium horse supplement

Vitamin E is vital for fertility and coat condition. It improves the immune response of the horse and is a natural antioxidant, helping to stabilize high fat rations. Selenium is required for bone and muscle integrity and metabolic functions. It is also an antioxidant.

In addition, this combination of vitamin E and selenium will improve stamina and performance, which are of paramount importance in sport and breeding horses. Ideally, 50 g per day per horse of vitamin E and selenium horse supplement are mixed in feed. The horses are then fed in the dry season up to the rainy season. This should be increased to 100 g per day per horse during the rainy season.

Liquid supplements

Liquid supplements (e.g. nutriquine) are formulated for horses and ponies, and may be added to the commercial horse feed ranges, providing the horse or pony with extra vitamins and minerals. They cannot be stored for long periods of time and should be bought as required. Some concentrated mineral and vitamin liquid supplements for horses provide a concentrated source of minerals and vitamins in a natural yeast base to supplement diets, which may contain low or inadequate levels of essential nutrients. Individual nutrient requirements of horses vary according to numerous seasonal conditions where sufficient amounts of minerals and vitamins may not be provided by feed alone for their optimal performance. As a result, liquid supplements cater for this need and enhances the individual performance of the horse.

Active ingredients of liquid supplements

1 kg of nutriquine contains:

Vitamin A	912 000 IU	Banox	50.0 mg
Vitamin D	536 000 IU	Organic chrome	92.0 mg
Vitamin B1	672.0 mg	Manganese	310 mg
Vitamin B2	608.0 mg	Zinc	3100 mg
Vitamin B6	357.0 mg	Copper	700 mg
Vitamin B12	1.6 mg	Cobalt	5.2 mg
Niacin	2232 mg	Iodine	7.2 mg
Pantothenic acid	964.0 mg	Iron	3120 mg
Folic acid	90.0 mg	Organic selenium	8.0 mg

Source: NRC 1989

The above constituents of liquid supplements are extremely beneficial to the horse in that they promote bone growth, optimal respiratory function, efficient digestion and amino acid metabolism, fecundity, higher milk yields, maximization of glucose metabolism and generally improve appetite.

Including natural yeast (namely Yea-Sacc 1026) into the formulation improves animal performance through greater roughage utilization by stimulating the growth of beneficial microbes in the hind gut (caecum) which accentuates the digestion of fibre.

By enhancing the gut function, there is an increase in forage intake and utilization, improved live weight gain and a reduction in metabolic and digestive problems.

Liquid supplements are, therefore, beneficial to the horse in that they improve all-round performance and condition.

4 General information about horse feed ingredients

Apart from the formulated feeds produced by stock feed companies, some companies also supply raw materials, which can be added to the daily diet of horses and ponies. These are:

a) *Wheat bran:* Bran provides the fibre and roughage content required by horses and ponies to create a good tone of the GIT.
b) *Hominy chop:* This is maize bran and is more of a carbohydrate than wheat bran.
c) *Crushed Maize:* Crushed maize is a good source of energy, especially for horses in hard training.
d) *Oat:* Oat feed is supplied as whole oat, which must be soaked in water for up to 4–5 hours before feeding. It may also be rolled, but then must be fed within 10 days, otherwise the nutritional value may be lost.
e) *Molasses:* Molasses is a cane based liquid supplement that makes the feed sweet, less dusty and more palatable.

Although the stock feed companies provide a well balanced diet for horses and ponies of all types, other requirements are needed to keep them in top condition, such as regular worming and teeth care.

Worming

Worming refers to the prevention and treatment of internal parasites. Foals should be wormed every 4 weeks with a broad spectrum remedy up to the age of 6 months. Thereafter they should be dosed for roundworms and strongyloides every three months. It is recommended that remedies be alternated to guard against the development of resistance against these dosing chemicals by the internal parasites. The treatment for liver fluke may be done at the beginning and end of each rainy season. All worming remedies should be administered as per manufacturer's directions.

Teeth care

If teeth are not rasped periodically (every 6 months), they become sharp and cause pain by lacerating the cheeks and tongue. If this pain persists, the horse will stop chewing properly and the nutritional value of the food will be lost, thus causing weight loss and decreased performance.

5 Guide to feed intake in horses

Table 8.1, shows the expected daily feed intake expressed as a percentage of the horse's body weight. The heavier the horse the more the feed intake expected. Feed consumption per day can therefore be calculated if the weight of the horse is known.

Table 8.1 Expected feed consumption by horses (% body weight)

	Forage	*Concentrate*	*Total Feed*
Mature horses			
Maintenance	1.5–2.0	0.0–0.5	1.5–2.0
Mares, late gestation	1.0–1.5	0.5–1.0	1.5–2.0
Mares, early lactation	1.0–2.0	1.0–2.0	1.0–3.0
Mares, late lactation	1.0–2.0	0.5–1.5	2.0–2.5
Working horse			
Light work	1.0–2.0	0.5–1.0	1.5–2.5
Moderate work	0.75–1.5	1.0–2.0	2.0–3.0
Young horse (foals)			
Nursing foal, 3 months	0.0	1.0–2.0	2.5–3.5
Weaning foal, 6 months	0.5–1.0	1.5–3.0	2.0–3.5
Yearling foal,12 months	1.0–1.5	1.0 –2.0	2.0–3.0
Long yearling, 18 months	1.0–1.5	1.0–1.5	2.0–2.5
Two year old, 24 months	1.0–1.5	1.0–1.5	1.75–2.5

From N.R.C. 1989.

6 Conclusion

Horse feeds are designed to supply the nutrients needed for promoting rapid growth, sound bone structure and good muscle development. Some feeds are specific to the needs of hard working and racehorses.

References

Araujo, K.V, 'Determination of apparent digestibility of nutrients of some concentrates and forages for equines using the mobile nylon bag technique', in Revista da SBZ, 1996, **25(5)**:944–956.

Mutetwa, L, 'Horse Feeding', in *SAFCO Quality Newsletter*, Vol. 2, 1999.

Nutrient requirements of horses Revised fifth edition, National Research Council, 1989

Slade L.M, 'Nitrogen metabolism in non-ruminant herbivores. The influence of non-protein nitrogen and protein quality on the nitrogen retention of adult mares', in *Journal of Animal Science,* 1970, **30**:753–760.

Woon Young Oh, tae Hong kang, Shin Heum Jim, Seong Koo Hong, Seung Ju Yang and Jae Hong Jung, 'Effect of feeding method on growth performance and carcass

characteristics of Cheju Nature Horses', in *Korean. Journal of. Animal. Science,* 1993, **35 (6)**: 505–509

9

RABBIT FEEDS

1 Introduction

Rabbits are normally raised for meat, as pets and for their fur. They are fun and easy to raise. They can use material that would normally be thrown away as feed ingredients and can convert this to meat and valuable fur. These could be carrot tops, sweet potato vines, bean leaves, green grass clippings and many kinds of foliage. Rabbits produce delicious meat. Their fur and feet are normally used to make toys and novelties. Manure from rabbits is high in nitrogen, such that it is very good for the garden and offers opportunities for use as ruminant feed, as is the case with poultry manure.

2 Rabbit feeds and feeding

Rabbits feed on colour and their chief protein sources are beans, peas, peanuts and the tops of these legumes. Some good sources are young grasses, cereal grains, rice bran and wheat bran. When ration formulation is done for rabbits, the crude protein levels that should be achieved for maximum performance are 12% for maintenance, 15% for pregnancy, 16% for growth and 17% for lactation.

Energy sources are typically cereal grains like sorghum, roots and tubers like sweet potatoes, cassava, and sugar cane. Leftovers like breadcrumbs can be used to fatten rabbits as long as the level of carbohydrates is controlled so that they do not accrue too much fat. Too much fat deposition produces inferior quality meat and predisposes the rabbit to heat stress.

Fibre content should be kept at about 15% to produce good tone of the gastro-intestinal tract. Lower fibre level normally causes diarrhoea, while too much fibre results in poor growth. Fibre for rabbits is typically supplied by the vegetable fraction. Roots and cereals are poor sources of fibre.

Minerals and vitamins are normally supplied in adequate amounts by the green feeds. However, some salt should be added at 0.5–1% level of total diet for appetite. Water provision is critical to rabbits as they can more easily die from lack of water than from lack of food. Consequently, plenty of clean water must be provided all the time. A doe that is suckling can drink up to 4.5 litres of water per day for maximum production.

The feeding of rabbits should ensure that they receive a variety of feed ingredients in their diet and not just one item. Commercial rabbit pellets are available and these constitute a completely balanced feed. However, over-shelved pellets deteriorate in their vitamin content and are not good for rabbits.

3 Feeding management of rabbits

When the diet of the rabbit has to be changed, the process must be done by gradual substitution over 3 to 5 days. Overfeeding of greens must be avoided due to their high water content. The problem is that the dry matter intake will be limited. Rabbits should be fed twice daily, in the morning and in the evening. They must be fed more in the evening as they are naturally more active during the night. The fresh feed should not be mouldy and should be kept away from rats and dogs as disease transmission can easily occur.

4 Conclusion

Feeding rabbits is very easy and is a very suitable commercial venture for smallholder farmers and resource poor farmers. Some farmers fail to make the best out of this enterprise by over dependence on swill (kitchen leftovers) and vegetables from their orchards. Farmers are advised that the other feed ingredients mentioned in this chapter are vital for high levels of production.

References

Commercial rabbit raising, Agricultural Information Bulletin Number 300, Washington, D.C, USA Government Printing Office

Cheeke, F.L, *Rabbit feeding and nutrition*, Orlando, Academic Press, 1987, pp. 82–83.

NRC, *Nutrient Requirements of Rabbits*, Washington, D.C, National Academy of Sciences, 1977

Prasad, R., SA Karim, and BC Patnayak, 'Growth performance of broiler rabbits maintained on diets with varying levels of energy and protein', in *World rabbit Science,* 1996, **4**:75–78.

Prasad, R., AK Tyagi, and SA Karim, 'Growth performance of broiler rabbits maintained on diets utilizing roughage and tree leaves', in *Indian Journal of Animal Nutritions*, 1996, **66**:835–838.

Raising Rabbits in Belize, Heifer Project International, Belmopam, National 4 – H Center, 1978

10

FEEDS FOR GAME ANIMALS

1 Introduction

The word *game* refers to wild animals that find themselves fenced into commercial farms and are then managed for commercial gain, in a manner that is environmentally friendly. Due to the increasing amount of game being kept for commercial purposes, in many countries stock feed producers have developed a range of feeds suitable for game kept on the veld.

2 Game cubes

This is a 16% crude protein cube with a high level of energy. It is suitable for partially zero grazed animals, animals newly introduced to an area and in areas where grazing and browsing are scarce. This is a critical feed source during the long tropical dry seasons.

3 Game getter block

This is a palatable low cost block. It is high in salt and molasses. This block is used for the purposes of attracting game to viewing areas or when some handling operation needs to be done. Game block (16% CP) is a high protein version of game feed. This is a protein block suitable for year round use. It contains no urea and, therefore, poses no danger of urea toxicity.

4 Phosphorus block/lick

Phosphorus can be provided in the form of a 30 kg block or loose lick. This supplement contains calcium and phosphorus that is needed for improved bone and horn growth. The same supplement is widely used in beef ranching. A good example is the Bosfos block that contains 7% phosphorus.

5 Browse plus lick

Browse plus lick is designed to assist browse utilization in harsh environmental conditions. It is believed to be very useful in situations where game relies on tanniniferrous foliage. Tannins bind to proteins and make them unavailable. Browse plus active ingredient, polyethylene glycol (PEG) binds to tannins and makes proteins available to the animal that would otherwise be complexed by the tannins. So browse plus is a rumen modifier, which can be included in any of the game feeds. It makes protein from high tannin and/or high lignin browse more available for digestion by binding the anti-nutritional factors (tannins). Browse plus should be used in winter or during a drought to maximize the intake of browse. It is most useful in tropical environments where game animals have to rely on foliage browsing during the late dry season when there is little or no veld grazing.

6 Rock salt

Rock salt is available in varying sizes for distribution in paddocks or near game viewing sites. It improves dry matter intake.

7 Game nuts

Game nuts are just as good as game block (16% CP), and this is a balanced feed aimed at game animals under intensive management systems, or with access to poor quality forages.

The recommended feeding rates for game nuts are:

- Impala – 0.5 kg nuts/head/day
- Sable, kudu – 1.0 kg nuts/head/day
- Eland – 1.5–2.0 nuts/head/day

A variety of medicinal feed additives can also be included in these nuts. These serve to allow preventative or curative treatment to the wildlife without subjecting the game to the stress and trauma of capturing techniques.

8 Wormicides

Game animals face many challenges, especially during the rainy season when internal parasite burden increases. Wormicides can, therefore, be incorporated into game nuts. Use of the wormicide has been shown to improve live weight gains, especially under intensive conditions. Best results occur if given four weeks after the first rains, pre-winter, or when animals are in a boma awaiting translocation. Feeding is recommended for ten days at the same intake rates recommended for game nuts before translocating the animals.

9 Flukicides

A flukicide can be included in the game nuts and is effective against immature and mature fluke. The ideal time for dosing is one month into the first steady rains. Dosing should be done again four to six weeks later for effectiveness. The recommended intake rates are identical to those for nuts, with the duration of feeding the medicated nuts being two days.

10 Crocodile complete feed

Crocodiles are carnivores and can be expensive to rear. Crocodile complete feed contains about 44% CP. This is a complete ration fed to crocodiles of all ages, without the need to feed meat. The feed comes in meal or nut form and contains a mineral and vitamin premix to meet the requirement of crocodiles. The nuts are fed as is and the meal is normally stuffed in sausage-skins to make it more palatable as the crocodile would perceive it as meat, due to appearance, colour and aroma.

11 Crocodile concentrate

A crocodile concentrate with slightly lower CP content (i.e. at 40% CP) is very suitable and normally fed from hatching to slaughter. The difference with the one discussed above is that it must be fed in conjunction with either chicken, beef or game meat. The feed also contains minerals and vitamins to meet the requirements of crocodiles. The mixing ratio is 60:40 (meat:concentrate) on weight basis, to achieve a complete balanced feed. The final mix can be fed as is, or in sausage skins as well. As a general guide to feed budgeting, it is helpful to remember that crocodiles consume between three to five per cent of their body weight per day.

12 Conclusion

Game farming has become fashionable and is now a brisk business throughout the world. It now calls for deeper knowledge of game nutrition, as these animals are now kept in confinement and in large numbers naturally unsustainable within the limited space. Therefore, principles of zero grazing must be applied as in domestic intensive livestock production.

References

Church, DC, ed., 'The nutrition and feeding of captive ruminants in Zoos', by Wallach, J.D. in *Digestive Physiology and Nutrition of Ruminants. Volume 3 – Practical Nutrition',* The OSU. Bookstores, Inc., Corvallis, Oregon, USA. 1972, pp. 292–307.

Denholm, L.J, 'The nutrition of farmed deer', in *Deer Refresher Course,* 1984, **72**:662–707.

Hudson, R.J. KR Drew, and LM Baskin,. *Wildlife production systems*, Cambridge Cambridge University Press, 1989, pp. 1–72.

Robbins, C.T, ed. *Wildlife feeding and nutrition,* second edition, San Diego, Academic Press 1993, pp. 1–19.
Wohlbier, W. 'Supplementary feeding of game animals and fowl' in *Roche Publication Index,* 1974, pp. 14–63.

11

LIVESTOCK FEED ADDITIVES

1 Introduction

Feed additives are high value feed ingredients that are included in the final diet for specific reasons meant to boost animal performance. Animal feed companies manufacture a wide choice of premixes which complement their stock feeds range. In addition to the maxi-packs which contain both premixes and macro-elements, stock feeds firms strive to make a commitment to quality control so as to ensure that farmers receive products that guarantee the improvement of animal performance.

2 Types of feed additives

Most animal feed companies offer farmers the following five star service:

(i) They manufacture a complete range of premixes for all classes of livestock,

(ii) Only the best raw materials sourced from reputable manufacturers are used,

(iii) Standard premixes are manufactured to internationally recommended nutrient levels to ensure optimum livestock production when added to balanced feeds, either using locally grown and imported raw materials.

(iv) They provide a technical advisory back-up service to all customers through a team of qualified and experienced nutritionists or technical company representatives, and

(v) In addition, animal feed producers manufacture custom premixes of vitamins/minerals to meet specific customer needs and also include feed additives that are difficult to obtain in the form of raw materials for home mixers.

Dairy feed additives

Common additives used in the dairy industry target high producers mainly. However, some additives also exist for low to medium producers. In most cases, the additives come on the market as various trade codes like DS or DI. Normally, 1 kg of dairy feed additives treats one metric tonne of feed.

Maxi-packs are normally meant to produce maximum effect, or make the animal display its maximum potential in as far as production is concerned. Examples of maxi-packs, like DL and DH, are also available for lactating cows and heifers respectively. In addition, specific feed additives and buffers can be added to enhance growth and milk production. The most widely used rumen buffer is limestone flour. Its main function is to buffer the rumen pH, especially when highly rumen degradable and fermentable diets are on offer to the animals. The normal rumen pH should be maintained between 6.65 and 6.69.

Fish feed additives

Premixes for fish, normally produced in packs of 9 kg and/or 4 kg are suitable for efficient fish feeding. Some are good with trouts, while some are geared for tilapia breeds.

Pig feed additives

Different pig feed additives target the different classes of pigs, namely creep, sow and boar and grower/finisher categories. Normally, 3.5 kg of the additive treat 1 metric tonne of feed.

Maxi-packs are also available and compounded in the final diet at a rate of 50kg/unit/MT. Maxi-packs for pigs can be targeted for creep feed, for sow and boar meal, grower meal and for finisher meal.

Some pig premixes and maxi-packs are also available and these include animal protein. Regulations governing the feeding of animal products to certain classes of livestock should be observed in particular countries or trade groupings. For countries exporting animal products to the European Union for example, farmers should not use feeds from animal products such as blood meal, meat and bone meal, to feed ruminant animals.

Poultry feed additives

Premixes for poultry (i.e. for broilers and for layers) are used extensively in the poultry industry in the following rates: 3kg for broilers, 2kg for layer per metric tonne of feed. Most animal feed companies can offer custom-made premixes and maxi-packs specifically formulated for poultry requirements.

Ostrich feed additives

Common feed additives for ostrich feeds are also made for either starters or growers, and for breeders. As a general guide 6.5 kg of ostrich feed additives treat 1 metric tonne of feed.

3 Vitamin C (ascorbic acid) as a feed additive for optimum performance

Vitamin C is ascorbic acid. It is vital for optimum performance of various livestock. Vitamin C in the feed has many functions, of which the following are key:

- • It alleviates stress,
- • It enhances resistance to diseases,
- • It improves egg quality, and
- • It increases reproductive performance.

Vitamin C is a compound synthesized from glucose. Domestic birds are generally capable of biosynthesizing vitamin C.

However, research over the past 10-15 years suggests that modern high performance birds are not able to synthesize ascorbic acid in quantities sufficient to meet the physiological needs for optimum performance. Field data and experimental results have shown that supplementation with vitamin C is, therefore, necessary for optimum growth, egg production, fertility, et cetera. The involvement of vitamin C in physiological and performance parameters points to the fact that vitamin C enhances growth performance and liveability of chicks. It also has been found to boost the natural defence mechanism (immune response) and, therefore, is a stress reliever. It has also been found to have toxin neutralization properties. In layers, vitamin C consistently improves fertility of the hens and hatchability. The latter is sustained through optimum eggshell and bone formation. It is now well proven that supplementation is especially important when birds are suffering from environmental, nutritional, pathogenic or other severe stressors.

In toxin neutralization as mentioned above, vitamin C acts as an antioxidant by inactivating the highly reactive free radicals, which are associated with damage to both intra- and extra-cellular structures. Vitamin C is also involved in the regeneration of vitamin E, the activation of vitamin D_3 and the formation of various hormones.

Because of highly intensified production in modern poultry management, stressors such as crowding, transport, growth and heat cannot be avoided. Stress, which is the physiological response of the bird to stressors, causes body reserves to be transferred from productive functions, such as growth, reproduction and performance, to non-productive functions, for example thermoregulation, sweating and blood flow.

During stress, the hormone corticosterone is secreted by the pituitary gland, which controls the mobilization of body reserves. Vitamin C plays an important role in the management of stress by modulating corticosterone secretion. Under conditions of prolonged stress, vitamin C levels in the plasma fall, allowing an uncontrolled release of corticosterone, and subsequently a reduction in performance. This condition can be alleviated by supplementary vitamin C, as a feed additive. This enables the birds to maintain productive performance better despite stress, and to reduce stress-related mortality.

The integrity of skin and epithelial tissue, of which collagen is the principal component, form an effective barrier against penetration by pathogens. Since vitamin C is imperative for the proper synthesis of collagen, vitamin C deficiency compromises membrane integrity, allowing pathogens to enter and cause all sorts of infections.

There is considerable evidence that vitamin C has a direct viricidal and bactericidal effect, resulting in decreased risk of infection. Cells of the immune system contain high concentrations of vitamin C, necessary for antimicrobial activity. Since microbial infections have been consistently shown to increase requirements for vitamin C over and above the birds' biosynthesis rate, sufficient supplementation is required.

Vitamin C is directly involved in the conversion of vitamin D_3 to its active metabolite 1.25 (dihydroxy) cholecalciferol. The activated vitamin D_3 is essential for optimum active absorption of calcium and phosphorus from the intestine, and for the mineralisation of bones and the eggshell during critical deficiencies.

Recent research has shown that vitamin C supplementation is required for younger and older animals alike. This is for optimal calcium homeostasis which results in healthy bone metabolism, normal egg production and adequate shell quality. Several observations have shown that incremental vitamin C supplementation results in incremental egg production and a corresponding pattern of decreased egg breakages.

Another important observation borne out of many years of research is that vitamin C is directly involved in the synthesis of the sexual hormones. In breeders, the biosynthesis of vitamin C seems to be insufficient for optimum reproductive performance. Supplementation of male breeder birds with vitamin C increases sperm production and improves sperm quality.

It has also been seen that in layers, fertility and hatchability of eggs are influenced by a proper structure of the egg shell, allowing good respiration and exchange of gases, and prevention of infections by micro-organisms. Vitamin C supplementation has been shown to improve fertility, hatching and egg-specific gravity.

Vitamin C is known to play a vital role in intracellular detoxification systems. It helps to metabolise and excrete drugs, environmental pollutants and mycotoxins. In those environments where salty water poses problems, recent research has provided evidence that the detrimental effects of saline drinking water can be significantly lowered by using supplemental vitamin C as a feed additive.

Other important research observations are that vitamin C is able to counteract the detrimental effects on egg quality by environmental pollutants, like heavy metals such as vanadium and cadmium. Mycotoxins can be a plague to the poultry farmer since they reduce growth and egg performance and increase susceptibility to diseases. Dietary supplementation with vitamin C can help to metabolise the mycotoxins and minimize the toxin-related detrimental effects. The next chapter deals with mycotoxins in more detail.

Supplementary rates

Table 11.1 shows the recommended supplementary rates for vitamin C in poultry feeding.

Table 11.1 Vitamin C recommended inclusion rates in poultry

Poultry type	*Vitamin C (g/T feed)*
Chicks, starting	105–145
Chicks, growing/replacement	105–145
Broilers	105–145
Hens, laying	105–180
Hens, breeding	105–180

4 Conclusion

Feed additives are critical in that they improve the efficiency of feed utilization. As a result they allow animals to achieve or nearly achieve their natural genetic potential. This leads to higher yields, hence profitable livestock farming. In simple terms, feed additives allow ordinary animals to produce extraordinary results.

References

Hattori, K. S, and R Fukumi, 'The effect of saccharide additives on the fermentative quality of silage', in *Journal of Japan Grassland Science*, 1993, **39**:326–333.

Tan, S.H., DV Thomas, BJ Camden, IT Kadim, PCH Morel, and JR Pluske,. 'Improving the nutritive value of full-fat rice bran for broiler chickens using lipase-based enzyme preparations', in *Asian-Australian Journal of Animal S*cience, 2000, **13(3)**: 360-368.

12

MYCOTOXIN CONTAMINATED FEEDS

1 Introduction

Mycotoxin contamination is a global issue affecting crops in the field, in storage, and during mixing and delivery. These are toxins produced by fungal infestation of feed material in the field and are a natural occurrence as crops grow. Clays and diatomaceous earth products have traditionally been used to bind mycotoxins in feed. However, these products bind only a few types of mycotoxins, have high inclusion rates, and have been shown to bind minerals and other nutrients, too.

2 Effects of mycotoxins on livestock

Animals are affected by mycotoxins in many different ways. The common effects are:

- In dairy cattle and swine, reproductive disorders are common, signposted by swollen red vulva, vaginal and rectal prolapse. Animals also display reduced dry matter intake in a very evident way.
- Animals may show skin and gastrointestinal irritation, neurotoxicity (poor coordination). They may reproduce abnormal offspring and are more sensitive to diseases. Haemorrhages can also occur.
- Animal displays nervous system disorders, tremors, convulsions, diarrhoea, and necrosis of the extremities (gangrene). Abortion, stillbirth and agalactia (poor milk production) incidences can increase in a herd. In poultry, the most common symptoms of mycotoxicity are the blackening of the comb, toes and beak and this can manifest

in a group of animals. Liver damage can also occur in a herd.

- In pigs, feed intake drops and weight gain is static or marginal, with cases of induced vomiting. Vomiting and feed refusal at high concentrations of vomitoxins are evident.

3 Mycotoxin absorbants

Mycotoxin absorbants (e.g. Mycosorb®) is a unique formulation of esterified glucomannans (EGM) derived from the cell wall of the yeast *Saccharomyces cerevisiae.* This has been found to bind a wide variety of mycotoxins at low inclusion rates and does not bind vitamins and minerals.

The most unique attribute of the mycotoxin absorbant is its structure. The structure has an extremely high surface area available to bind several types of toxins. One gram of a mycotoxin absorbant, for instance, has 3×10^{10} particles. 500 grams of Mycosorb have been known to have a surface area of approximately one hectare (100m x 100m). A piglet eating one kilogram of feed, therefore, receives 2 hectares of a mycotoxin absorbant in terms of surface area. The large surface area results in significant mycotoxin absorption.

Studies from the Unites States, Canada, South America and Asia confirm the ability of mycotoxin absorbant to bind a broad spectrum of mycotoxins. The strongest binding occurs with aflatoxins, and significant binding occurs with zearalenone, citrinin, fumonisin and ochratoxin.

Binding sites of other absorbants are often occupied by other feed constituents, thereby reducing toxin adsorption. In comparison, studies have shown that a mycotoxin absorbant retained its ability to bind mycotoxins when mixed with poultry feed.

One of the valuable characteristics of a mycotoxin absorbant is that its toxin adsorption capacity is unaffected by pH changes. To be effective, binding capacity must be maintained across changes in intestinal pH. Mycotoxin absorbent's binding capacity remains stable as pH changes from 4.5 to 6.8.

The other key property of these absorbants is that digestive enzymes do not affect their activity. During digestion, the absorbant-toxin bond is exposed to attack by gastric and intestinal enzymes. Even after a two-stage incubation with pepsin, the absorbant retains the ability to maintain mycotoxins in the bound state.

Most mycotoxin absorbants are included in feed at 0.5 to 1.0 kg per tonne. When compared to aluminosilicate clays (commercial), mycotoxin absorbants are effective at much lower inclusion levels because of the vast surface area of the product. Research has shown that effective binding occurs at one eighth (1/8) the inclusion level of silicate clays.

The strength of binding toxins is so important in preventing toxicity. Comparing mycotoxin absorbent's ability to bind to various toxins, Canadian work has shown that not only does an absorbant bind a range of toxins, but also this binding is strong. In other words, the toxins are not easily released, offering effective protection to the animal.

One of the many causes of liver failure is over accumulation of toxins. For example, high levels of aflatoxin B_1 in layers' feed result in higher levels of the toxin in the liver. Mycotoxin absorbants, when added to the diet, reduce liver mycotoxin content in layers' feed by more than 50%. Egg production is, therefore, enhanced.

Mycotoxin absorbant inclusion rates

Commercial absorbants are low inclusion, all natural, nutritional mycotoxin binders for all species. The following inclusion rates are the standard recommendations. (table 12.1)

Table 12.1 Mycotoxin binder use rates in poultry, pigs and dairy cattle.

Poultry	
Breeding stock	1–2 kg/tonne of feed
Broilers	0.5–1 kg/tonne of feed
Layers	0.5–kg/tonne of feed
Pigs	
Sows	1–2 kg/tonne of feed
Grower	1–2 kg/tonne of feed
Finisher	0.5–1 kg/tonne of feed
Dairy cattle	
Per head	10–15 g/day
TMR	0.5 kg/tonne of feed

4 Moulds in livestock feeds

The common problems created by moulding are production of toxins, reduction of feed nutritive value, decreased palatability, increases in feed moisture and feed caking.

Moulds produce a large number of toxins, which result in reduced livestock performance, poor feed conversion and instances of livestock poisoning. Poisoning normally results due to the liver being overwhelmed by the toxin load. The liver fails to detoxify and the animal is poisoned.

Moulds use feed nutrients that could have been used by the animal for growth. Mould growth reduces feed energy content, lowers feed vitamin levels, and can damage feed protein and amino acids. The resultant effect is a lowered feed nutritive value. For instance, metabolisable energy loss in corn/maize due to moulding can be as high as 25%.

Even small amounts of mould growth create dust in feed, with off flavours and a musty odour contributing to poor consumption. Therefore, moulding reduces the palatability of feed and decreases feed intake.

Feed stored with the potential to mould often has above desirable moisture content (12%), and this problem is compounded by metabolic water produced during mould growth. The increased moisture content and mycelia mould growth cause feed to clump, creating handling problems due to caking of feed in storage containers.

Feed management and mould conditions

Moisture

Moisture is the single most important factor determining whether or not mould will grow. If moisture is reduced to 13–14%, mould growth will stop, but mould spores are not killed. Small changes in humidity can quickly bring feed moisture above the safe level. This is why a good mould inhibitor is critical for stored feeds and grains.

Temperature

Moulds are able to grow over a wide range of temperatures from less than 0°C (32°F) to over 49°C (120°F). The optimum temperature for most moulds is between 21° and 33°C (70° and 90°F).

Grain damage

Intact grain has a natural protective coat, which helps prevent mould invasion. On the other hand, broken kernels or insect damaged kernels are excellent sites for initiation of mould growth.

Drought

Grain produced during a drought is more likely to have insect and bird damage. This allows for greater infection by mould spores in the field. Extensive sampling has shown that moulds and their toxins are more likely during drought years.

Mould controllers for feed and stored grain

The control of moulds and their mycotoxins in animal feeds and feed ingredients continues to concern livestock producers and nutritional biochemists due to their damaging effects. One of the easiest and most economical ways to protect feed and grain against mould growth is by using chemical mould inhibitors. Several different kinds of chemicals have been used over the years (e.g. Mould-ZAP®). Factors affecting use are relative efficacy, cost and the legal status of the compound. Most commercial mould inhibitors are based on organic acids (propionic, benzoic, acetic, tartaric, and sorbic acids) and their calcium, potassium, and sodium salts.

Mould inhibitors are blends of organic acids and consist predominantly of buffered propionic acid. Propionic acid is widely recognized as one of the most effective mould inhibitors available.

Most mould inhibitors like sorbic acid have a very broad spectrum of activity against yeast, and moulds, but are less active against bacteria. Sorbic acid maintains its anti-microbial activity up to pH 6.5 but, as with other weak acid microbial inhibitors, its activity increases as the pH of the feed decreases. Sorbic acid is not toxic to animals and is thought to be metabolised like any other fatty acid. Sorbic acid, when used alone as an antimicrobial agent in feedstuffs, is normally applied from 0.02% to 0.1 5% by weight.

Each of the organic acids mentioned (acetic, benzoic, propionic, sorbic) have advantages and disadvantages in terms of their efficacy against yeasts, bacteria and moulds, range or organisms inhibited, ease of handling and costs. Used together, these organic acids retain their individual advantages and compensate for the minor disadvantages of each acid when used alone.

Mechanism of inhibition

Organic acids (such as propionic acid, acetic acid, sorbic acid and benzoic acid) inhibit micro-organisms by entering the cell in the undissociated form and then dissociating within the cell. This causes acidification of the cytoplasm and results in the inhibition of nutrient transport. The micro-organism's energy consumption increases in an attempt to maintain pH homeostasis. Because the organism's living processes occur over a relatively narrow pH range, the cell ultimately dies or is inhibited.

Most moulds produce significant amounts of carbon dioxide gas (CO_2) as they metabolise in feed. By measuring the rate of CO_2 production in closed feed samples, one can estimate the efficacy of mould inhibiting compounds.

Properties of broad spectrum mould inhibiting agents

Broad spectrum mould inhibitors contain a blend of organic acids, each with its own characteristics. For example, acetic acid is a colourless liquid which solidifies at 17°C (62°F) and is mixable with water. Acetic acid is more effective against yeasts and bacteria in feed and less active against mould. However, when acetic acid is applied at high levels (0.40 %–1.0%), it is effective against moulds in feed. It has higher antimicrobial activity as the pH is lowered, thus increasing the quantity of undissociated acid. Under regulations of many countries, acetic acid is generally recognized as safe for use in feeds.

The other useful mould inhibitor is benzoic acid. This preservative is often used in the form of sodium salt in white powder or flake form, and as an antimicrobial for foods. The optimum pH range for benzoic acid is 2.5 to 4.0, which is lower than the other acids found in mould inhibitors. Benzoic acid is highly effective against yeast and bacteria and, however, less effective at controlling moulds.

Propionic acid, in buffered form, is the major active ingredient in mould inhibitors. Straight propionic acid is a clear liquid, which is more highly corrosive than other mould inhibitors. Propionic acid also has a strong, irritating odour. Propionates are more active against moulds than benzoic acid, but have little activity against yeast. Industrially, propionic acid is widely used as a mould inhibitor in bread and cheese, and can also function as an antibacterial when used at higher levels (0.4% in animal feeds).

5 Conclusion

Mycotoxin contamination of feeds of livestock is common in hays as well as silages. The degree of contamination is affected by herbage growing and harvesting conditions, with wet weather during harvesting hay or crops increasing the probability of contamination. High concentrations of mycotoxins absorbed in body systems of livestock lead to depressed growth, lower feed efficiency, impaired immunity and decreased reproduction.

References

Botha, C.J., T.S Kellerman, and N Fourie, 'A tremorgenic mycotoxicosis in cattle caused by Paspalum distichum (L) infected by Clariceps paspali', in *Journal of the South African Veterinarian Association*, 1996, **67**:36–37.

Journet, M., E Grenet, M.H Farce, M Theriez, and Demarquilly, eds, 'Plant toxins and mammalian herbivores: Co-evolutionary relationships and anti-nutritional effects', by Cheek, P.R. and R.T Palo, in *Recent developments in the nutrition of herbivore*. Proceedings of the 4th International Symposium on the nutrition of herbivores, Clermont – Ferrand, September 11–15, 1995.

Schneider, D.J., C.O Miles, I Gaithwaite, A Halderen, J.C Van Wessels, H.J Lategon, and A Van Halderen, 'First report of field outbreaks of ergot alkaloid toxicity in South Africa', in *Onderstepoort Journal of Veterinary Reearch,* 1996, **63**: 97–108.

Seifert, H.S.H., *Tropical Animal Health*, Dordrecht, Khiwer Academic Publishers, 1996, pp. 441–496.

Singleton, V.L. 'Naturally occurring food toxicants, phenolic substances of plant origin common in foods', in *Advances in food Research* 1981, **27**:149–242.

13

GENERAL PRINCIPLES OF LIVESTOCK FEEDING OPERATIONS

1 Introduction

The most important aspect of feeding livestock is the efficiency with which meat milk and eggs are produced. Less and less feed should be consumed for any given level of production. Ordinary animals must be fed so that they produce extraordinary production levels. Its is also vital to ensure that extraordinary animals do not produce ordinary production levels.

2 Feed utilization efficiency

Feed efficiency is measured as the amount of feed that is eaten to produce 1 kg of meat (live mass basis) or 1 litre of milk. Ideally, feed efficiency for broilers should be 1.8–2 kg of feed per kilogram of meat produced, for pigs 4:1 and for ruminants 7–8:1. Feeding management should always aim to improve feed efficiency with time. If for beef cattle, the feed conversion efficiency can be improved from 8 kg feed: 1 kg live weight gain to 6.5:1 (which is being achieved by some farmers in the developed world), 300 kg of feed per head (during the fattening period) is saved. If the feed costs US 65 cents per kg, this represents a saving of $195 per head, $1,950 for 10 animals and $19,500 for 100 animals during the same period; this is a huge saving.

3 Practical ways of managing feeding for efficiency

Setting realistic targets

The farmer should set realistic performance targets in relation to the quality of animals being fed and animal production levels so desired. The farmer should provide a ration which, as precisely as is possible, supplies the right amount of nutrients, in the right proportions, to meet those set targets. Given technical limitations, it is advisable to assess feeds produced from the farm or bought in. The feed ingredients must provide the cheapest source of nutrients, but not necessarily the cheapest feed ingredients.

Feed mixing

Inadequate mixing of the ingredients can defeat the whole purpose of providing the correct amounts of nutrients in the right proportions. Therefore it is necessary to ensure that the feed is mixed properly (i.e. uniformly) so that every mouthful is the same for all the animals feeding from the same trough. This has been the biggest benefit provided by mixer wagons. Proper mixing allows for a steady supply of balanced nutrients to the animal's body.

When animals are feeding, it pays to check dry matter intake (DMI). Animals eat the same DM irrespective of moisture content.

Litter management

Litter management is vital. Too much litter creates off flavours that can affect the animal's appetite. No animal is comfortable with the smell of its own excreta. Cattle, for example, must be able to lie down in an area free of mud or wet manure. (Animals only grow when they lie i.e. are asleep). In heavily littered environments they will try to avoid lying down as much as they can due to the discomfort, and their growth potential is interfered with. Mud or wet manure at knee high has been found to reduce production (e.g. growth, daily gain) by as much as 50%. At hock level, production goes down by

30%. Heavily littered environments breed parasites and this is expensive as insect control must be done vigorously.

Ration composition

Ration composition is very important. Farmers should not feed too much hay or too much concentrate or any other ingredients. For example, normally molasses should not exceed 8% in pen fattening beef feeds. Levels as high as 12–14% molasses can cause blind staggers in animals. This is a nervous problem caused by the softening of the brain tissues, especially if drinking water contains sulphates. However, lack of sulphates in the drinking water for cattle has seen, beef farmers feeding high levels of molasses in some countries. This is normally accompanied by feeding 3 g of vitamin B_1 (thiamin) per day per head.

Feeding too much of one ingredient causes gastrointestinal problems, e.g. bloat and acidosis in cattle or scourers (diarrhoea) in other species. Laminitis is caused by too much feeding of cereals. This is infection through the hooves. The infection results due to the softening of hooves starved of protein supply. Hooves are chemically proteinous and need a constant adequate supply of protein in order that their integrity is maintained. So, when a farmer mismatches protein supply by over supplying cereals, the hooves lose their hardness, become soft and infection occurs.

Simple observations can be made on the dung of cattle and this can help to diagnose feeding problems. Dung structure and texture are simple but helpful qualitative parameters. The bigger the dung pat the better. Semi-liquid dung (no bubbles) indicates acidosis. Dung colour is another important aspect of simple observations and the colour should be brownish yellow and not dark.

Cafeteria feeding

Cafeteria feeding is a situation were animals are allowed to eat as many feed ingredients as is possible. As a result the ration should contain as many feed ingredients as possible so that the diets are more palatable. A ration containing a variety of ingredients is more palatable and less monotonous to the animal. Animals are able to harvest a variety of chemicals from the varied array of ingredients and this translates into better tasting meat. So what it means is that it would be better to have more than one protein source (soya, sunflower, fish meal, et cetera) and also more than one energy source (maize meal, molasses, et cetera). Beef cattle produced under tropical grazing normally produce delicious meat because they feed on a variety of veld grass and forage that has been produced under natural photosynthesis. Zero grazed beef on a limited array of feed ingredients produces flat tasting meat.

Feed bunk management

Feed bunk management is a science and an art. The feed bunk should slope towards the animal so that it can clean the trough without over twisting its neck downwards, hence without difficulties. Feed should be offered less at a time, and not all at a time, to avoid wastage. The feed placed in the bunk should be uniform. This is because animals have favourite feeding spots and should get a uniform array of nutrients each time they visit their favourite feeding spots when hungry. In addition, the farmer should ensure that the feed in front of the animals is as fresh as is possible, and that feed is available for at least 22 hours per 24-hour period. Animals, just like human beings, do not like to eat a mixture of fresh feed with the previous day's refusals. The trough should be cleared of the previous feed and cleaned. The animals would also not prefer fresh feed fed in dirty troughs. One of the most important provisions when feeding animals is clean water. The farmer should make sure fresh water is available at all times. The water troughs should be cleaned regularly as well.

Animal activity during feeding should be monitored closely. On average, at any one time, 30% of the animals should be seen eating, 30% chewing, 30% lying down and 10% moving around, otherwise the farmer should investigate if a large number of animals are off feed.

4 Stress management during feeding

The farmer should also provide comfort in the feeding pens and avoid stressful situations like overcrowding. Cattle should typically be provided with 8–10m^2 of space per head. Overcrowding predisposes animals to respiratory problems. Other stressful situations that must be avoided include shortage of feeding space at the trough, heat stress, feed and water deprivation. Stress reduces immunity and causes viral infections. Viruses do not kill the animal but compromise immunity, and then opportunistic bacterial infections set in and do the lethal damage. Nutrients are wasted as the animal fights numerous infections.

Respiratory disease identification is vital. It pays to pull out sick animals quickly from the feeding pens and relocate them to hospital pens so that they can be fed separately, with close monitoring. They will also not be stressed by the healthy ones, through pushing and shoving in the feed bunks. Once the sick recover, they should then be moved to recovery pens, and fed separately long enough before they rejoin the rest. As a management tool, the farmer should record and provide the veterinarian details upon request pertaining to % morbidities, % mortalities (which should be less than 4–6%), and case fatalities.

5 Feeding management as value addition

Feeding is a commercial business, and more precision in the business is the secret to success and improved margins. The nutritional requirements for growth, egg production, fattening or milk production should be known. Once targets are set, the required nutrients must be applied in order to achieve these targets. This entails knowing the nutrient content of feeds being used on the farm, which means feed materials should be analysed for key chemical components like dry matter, crude protein, fibre, energy, minerals, vitamins and anti-nutritional factors. Failure to do this can cause huge differences between what the farmer thinks he or she is feeding and what the animals are actually getting.

6 Conclusion

Feeding management is an art and a science, and wisdom both passed on from predecessors or peers and/or learned are equally important. The farmer can have an extraordinary animal, provide it with extraordinary feed, but if feeding management is not tip-top, that animal will produce ordinary results.

References

Cattle Producers Association, 'Management of Beef Cattle' in *Beef Production Manual*, Zimbabwe Commercial Farmers Union, 1998, 98/2, pp. 1–15.

Grant, R.J. and J.L Albright, 'Feeding behaviour and management factors during transition period in dairy cattle', in *Journal of Animal Science*, 1995, **73**:2791–2803.

Morrison, F.B. 'Feeds and Feeding', twenty-second edition. Ithaca, Morrison Publishing Company, 1956, pp. 26 –39.

Preston, T.R, 'Fattening beef cattle on molasses in the tropics' in *World Animal Review* 1972, **1**:24

Topps, J.H. and J Oliver, 'Systems of feeding livestock: animal foods of Central Africa', revised edition, *Zimbabwe Agricultural Journal Technical Handbook,* 1993, **2**:8–147.

14

FEED FORMULATION TECHNIQUES

1 Introduction

Feed formulation is the process of creating formulae for combining the various feed ingredients. The formulae must permit the mixing of feed ingredients in proportions that will satisfy specific animal requirements in terms of energy, proteins, minerals, et cetera, or a combination. In other words, the formulae must produce a balanced diet. It is fairly easy to calculate the proportions of mixing a diet composed of two ingredients. This is because simultaneous equations can be used to solve the unknown. The moment three ingredients are involved, three equations have to be solved simultaneously. With four ingredients, four equations are constructed, and so on. We encourage farmers to do cafeteria feeding as this gives the animal the opportunity to feed on a more palatable and nutritionally useful diet, but this presents problems as complex equations have to be constructed and solved using the matrix method or linear models. These methods, when done manually, can take months to produce solutions, hence are tiresome and laborious. The modern day nutritional practitioner uses computer aided formulation techniques for speedy solution generation.

2 Information needed to formulate a ration

There are two important pieces of information that are needed to enable a nutritionist to formulate a ration or diet for a specific animal. These are the nutritional requirements of the animal concerned and the chemical properties or nutritive value of the feed ingredients that are to be used in formulating the diet.

Nutrient requirements of specific breeds and classes of animals depend on expected level of production: e.g. daily gain, lactation, pregnancy, wool, eggs production. This information can be calculated or obtained from feeding standard tables.

Nutritive value of the feedstuffs (ingredients) depends on the chemical attributes of the feed ingredient. The necessary information is obtained by proximate analysis or from tables on chemical composition published by relevant authorities, like the Agricultural Research Council, National Research Council and the International Research Association.

3 Methods of formulating rations

Various methods are used to calculate the proportions in which feed ingredients have to be blended in order to produce a balanced diet. The complexity is proportional to the number of ingredients that have to be included in the final formula.

Manual methods

The manual methods are based on empirical evidence, i.e. common indigenous knowledge that has been gleaned and generated over generations (from experience), and these have become average levels, or, as some may want to call them, usual levels. For instance, for fattening beef, the common diet would contain 70% grain, 20% roughage and 10% protein and this gives a final diet that contains 130 g/kg (13%) crude protein, and 11.5 MJ/kg of metabolisable energy. Also included in this formula is limestone flour, rock salt, monocalcium phosphate at 0.5% inclusion rate in each case. As you can see from the above, the final formulation contains 101.5% proportions in total. It doesn't really matter or make the diet unbalanced. This is what makes it an average formula. Even with situations were the total proportion is formulated to 100% exact, you will find that, as the mixing is done in the practical sense, there are small errors in weighing that can in reality produce a final proportion of 99%, or slightly more than 100%. These small errors in either direction are insignificant to the quality of the diet and are actually normal.

This method can be improved by first analysing the ingredients for protein and energy. Then a quick check is instituted where nutrients provided by each ingredient are calculated so that their sum per kilogram gives crude protein level of 13% and metabolisable energy level of 11.5 MJ/kg. Adjustments, where necessary, are done by substituting one ingredient with the other until the correct diet is arrived at.

For pigs, the usual or average creep diet (for piglets) from empirical evidence contains 64.3% polished maize meal, 31.0% soya bean meal, plus mineral/vitamins mix, lysine and tryptophan to make it 100%. For growers the formula is 70% maize meal, 23% soya bean meal plus mineral/vitamin mix, lysine and tryptophan.

For smallholder pig farmers, this formula is simplified for ease of administration to 3 parts maize mixed with 2 parts soya bean meal. This approximately gives a final diet that contains 167 g/kg (16.7%) crude protein and 11 MJ/kg of metabolisable energy.

For dairy cows (producing 20 kg of milk per cow per day), a diet must be formulated that contains at least 130 g/kg (13%) crude protein and more than 10.5 MJ/kg metabolisable energy. On average, this can be achieved by the following formula: 35% maize silage, 13% good quality hay, 39% maize meal, 13% soya bean meal or cotton seed cake, or a combination of the two, 0.5% limestone flour and 0.5% rock salt.

For calves, a colostrum substitute can be made. If a cow dies before her calf received colostrum, the farmer should get some quickly from another cow, or formulate a substitute as follows:

- 1 fresh egg
- 1 litre clean water (warm)
- 1 teaspoonful cod liver oil

- 3 teaspoonfuls of castor oil

This should be mixed well and fed to the calf three times per day for the first four days, then continue with regular milk.

For poultry, the following table shows the nutrient requirement and the average formulae.

Table 14.1 Nutrient requirements for poultry

	Layers	*Broilers*
MJ/kg	11.5–12.5	12.5–13
CP %	16.5–17.5	16.5–20
Ca %	2.3–3.5	0.5
P %	0.6–1.0	0.3
The exact nutrient requirements for different classes of poultry are given as:		
Chick mesh	20% CP	12 MJ/kg ME
Starter	18% CP	12.5 "
Growers	17% CP	12.8 "
Finisher	16% CP	13 "

The likely rations for poultry (i.e. kilograms of ingredient per tonne) are given in table 5.2.

Generally, the approximate formula for mixing ingredients for a particular animal can be obtained by directly observing the animal as it eats in a natural set up. Records are made judiciously on the preferred ingredients, how long they spend on these per day, and how much of each ingredient they take in per day. The individual ingredients are then mixed in the approximate quantities they have been eaten by the animal per day over a two or three week period and then ingredients are analysed individually and as a compounded diet. Properly designed and controlled experiments can then be done to validate this data. This will lead to the generation of optimum formulae for the various physiological states and production levels.

Pearson square method

This method is more accurate than just using average diets but only works where two ingredients are used, or where 3 ingredients are used with the level of one of them fixed at a particular rate. Ingredients can also be grouped into two groups and the composite portions analysed, and the information is then used in the Pearson square method. The chemical composition and nutrient requirements to be satisfied should be known.

For example, if maize meal and soya bean meal (SBM) are to be used to formulate a diet for laying birds, the diet to be formulated should finally contain 160 g CP/kg. If maize contains 90 g/kg crude protein (CP) and SBM contains 360g/kg CP, calculate the mixing ratio (formula) of maize to soya bean meal.

Solution: construct a Pearson square by simply criss-crossing lines as shown below.

On the left hand side, insert the ingredients and the nutritive value in terms of CP. Put the target CP content of the final diet at the centre or point where the lines cross each other. Then subtract diagonally and insert the answer (being the final proportion of ingredient in the same row) where that diagonal line ends as shown in the following diagram:

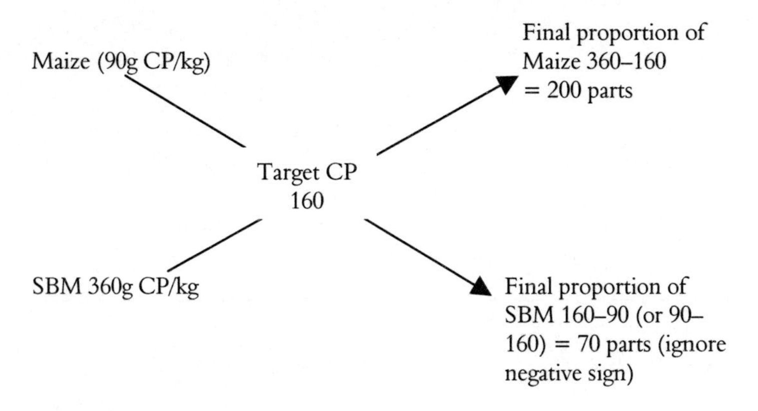

So, the formula is 200 parts maize mixed with 70 parts soya bean meal in order to produce a diet that contains 160 g/kg (16%) crude protein. This can be reduced to proportions as a percentage for simplicity and clarity as follows:

$$\textit{maize} = \frac{200 \text{ parts}}{200 + 70 \text{ (total parts)}} \times 100 = 74\%$$

$$\textit{SBM} = \frac{70 \text{ parts}}{270 \text{ (total parts)}} \times 100 = 26\%$$

This method can be used with any nutrient as the reference point (i.e. energy or fibre content.) However the parameter crude protein is favoured because for any particular class of animal or species, if you balance for proteins, you are rest assured that energy is also balanced for. It's only in a few cases, as with poultry (which feed to satisfy their energy requirement), that extra oil or fat might need to be added to correct the energy deficit.

Algebraic summation method

For those who are mathematically minded, the above solution obtained with the Pearson Method can be obtained using algebra. Let x % be the proportion of the ingredient SBM, y % be the proportion of maize and [CP] the concentration of crude protein.

Since x % + y % =100%, we can say x + y = 1. Therefore y = x – 1.

X and y are the proportions of the ingredients. Now we can find out the proportions of crude protein that is contributed by SBM and maize to give the proportion (%) of crude protein in the final diet. This is not obtained by simply adding the percentages of crude protein in SBM and maize, but by first calculating the contributions of each. SBM will contribute x multiplied by the concentration of crude protein in SBM i.e. x[CP]. So, the concentration of crude protein in one kilogram of the final mix (the diet or the ration) is equal to the contribution from SBM, plus the contribution from maize.

Therefore:

[CP] in 1 kg of final mix/diet or ration = (x)([CP] of SBM) + (1 – x)([CP] of maize).

Note that the concentration of CP in 1 kg of the final mix is just as good as the percentage content of CP in the final diet. 1 kg is used as a reference point for clarity purposes.

If we rearrange we get:

$$x = \frac{([CP]\ of\ diet) - ([CP]\ of\ maize)}{([CP]\ of\ SBM) - ([CP]\ of\ maize)}$$

$$x = \frac{(160 - 90)}{(360 - 90)} = (0.26)$$

$$\therefore\quad y = (1\text{-}x) = 1\text{-}0.26 = 0.74$$

Therefore the inclusion levels are 26% and 74% for SBM and maize respectively.

Computer Aided Formulation

Since the inclusion of more ingredients in a ration is highly recommendable but presents difficulties in as far as solving complex equations is concerned, computer programs that utilizes linear modelling techniques are in use now and make operations easy and fast. Packages exist for demonstration purposes as downloads on the Internet. There are also packages for training purposes as well as huge programs for industrial stock feed manufacturing. Smaller packages also exist that can be bought and used easily by farmers. All these programs are user friendly

4 Conclusion

Feed formulation is critical to successful animal production. Gone are the days when animals could just be let loose and expected to fend for themselves. The modern day animal producer must produce animal products quick enough and as efficiently as is possible. This can only ensue if the animals are given balanced diets. Grazing lands are becoming less and less in hectarage while, demand for animal products is skyrocketing. Feed formulation is therefore an essential component in the livestock industry.

References

De-xun, L., X Benhai, and Y Jing, 'A new technique for formulating protein supplements for ruminants', *Proceedings of the eighth World Conference on Animal Production,* June 28–July 4, Seoul National University, 1998 PP. 10–11.

15

IMPROVING THE UTILISATION OF LIVESTOCK FEED RESOURCES

1 Introduction

The utilization of feed resources by livestock is a summation of intake, digestion (plus rumen fermentation and hindgut fermentation in ruminants), absorption, translocation, and metabolism of nutrients. Metabolism involves catabolism (breaking down into the smallest building blocks (molecules) and anabolism (i.e. the formation of new, functional large molecules). Metabolism occurs at a defined pH, osmotic potential and, in the presence of appropriate minerals, vitamins, hormones and enzymes. The result of metabolism is the provision of energy and proteins for the maintenance of body weight and functionality of tissues or repair of worn out tissues, or production of more tissues for growth, egg production, milk production, wool, et cetera. Utilization of nutrients for production (i.e. over and above maintenance requirements) is the profit margin a farmer aims to optimise. This optimisation can only come about if the utilization of feed resources is improved.

When feed resources are utilized efficiently:

- Satiety is achieved,
- Digestion is optimised,
- Absorption and translocation processes are optimised, and
- Redox (metabolism) reactions are encouraged forward and optimised.

Every effort should be made, therefore, to improve the physical and chemical properties of feed ingredients in a bid to optimise utilization. The problems that are associated with feed utilization are:

- Malnutrition
- Limitations due to type of feed (e.g., crop residues versus natural pastures).
- Seasonal changes (fluctuating feed supply and quality).
- Animal welfare (i.e. stress, housing conditions, handling and social interaction, breed, disease, accessibility, presentation, frequency, refusal allowance, refusal removal).
- Feed palatability (i.e. flavour, aroma, pH, toxicity, water content, and previous experience).
- Physical properties of feed, i.e. particle size, volume and density, morphology (leaf to stem ratio), plant species (legumes versus grass).
- Nutrient availability, i.e. imbalances, fermentability and end products.
- Feed digestibility. A 1% increase in digestibility results in 9% higher feed efficiency and results in 10% increase in milk yield in dairy animals. This represents important economic gains.

2 Ways of improving utilization of feed resources by livestock

As has been alluded to in Chapter 13, feeding of heterogeneous rations is more advantageous compared to mono or two ingredient rations. A meal that is made up of a variety of ingredients improves palatability and, therefore, intake. The feed has desirable organoleptic properties which entice the animal to consume more of the diet offered. These organoleptic properties are finally conferred to the products, i.e. meat, eggs and milk. As a result, the products so produced (meat, milk, eggs) are tastier and favoured by the consumers.

Appropriate technologies used to improve feed utilisability include:

- Industrial oil expression/extrusion of soya seeds,
- Heat treatment (boiling, roasting),
- Milling/pounding, grinding,
- Solar treatment (under experimentation in Southern Africa),
- Sprouting soy beans,
- Soaking (hydrating crop residues),
- Ammonisation with 4% solution of urea (see Chapter 18 item 3),
- Biotechnology (silage making); use of mixed silage grasses and legume silage,
- Sodium hydroxide treatment,
- Microbial treatment (yeast and white rot fungi). Hoyos *et al.* (1987) have demonstrated that yeast culture improved milk production from 26.0 kg per cow per day to 30.4 kg in high yielders and 18.6 kg to 20.3 kg in medium yielders. Another trial managed to achieve increases from 31.6 kg to 34.6 kg with yeast cultures seeded in dairy cow feeds,
- Use of enzymes, e.g. cellulase,
- Pelleting (removes dustiness, and enhances handling),
- Compressing feed into cubes and blocks,
- Judicious mixing (ration formulation),
- Inclusion of limestone flour in ration to buffer rumen pH so that rumen fermentation occurs optimally,
- Synchronizing supply of nutrients to the animal, e.g. soya cake and maize meal (or urea and molasses) so that there is coordinated, coupled fermentation in the rumen,
- Inclusion of salt (sodium chloride) to enhance appetite,
- Feeding according to animal needs, especially in dairying and beef production,
- Introducing new forage crops which are leafy, fast growing and low in anti-nutritional factors,

- Use of by pass protein using blood meal, for example, or condensed tannins,
- Diluting anti-nutritional factors, e.g. restricting cotton seed cake inclusion rate to a maximum of 5% so that the anti-nutritional factor gossypol is diluted.

3 Conclusion

Feed offered to the animal must be effectively utilized. That is, the animal must benefit from the diets as much as is possible at the cellular level. Feed must therefore be easy to digest, absorb, transport and metabolise. These processes must occur as efficiently as is possible, meaning the animal must produce more and more animal products from less and less dietary nutrients.

References

Emmert, J.L. and D.H Baker, 'Use of the meal protein concept for precision formulation of amino acid levels in broiler diets', in Journal of Applied Poultry, 1997, **6**:462–464.

Liu, T.Y. and W Cheng, 'The study on the ammoniated straw technology for efficiently fattening beef cattle', *Proceedings of the eighth World Conference on Animal Production,* Seoul National University, June 28–July 4, 1998, pp.78–80.

Shrestha, H.R., B.S Kuwar, P Mandal, M.S Thapa and S.B Panday, 'Effect of feeding urea and molasses treated rice and wheat straw diet on body weight gain and carcass characteristics of male calves', *Proceedings of the eighth World Conference on Animal Production*, 1998, pp. 70–71.

16

BIOTECHNOLOGY IN THE FEED INDUSTRY

1 Introduction

Biotechnology (in the feed industry) is the use of feed additives derived from living organisms in the form of natural substances like yeasts, bacteria, enzymes, plant substances, et cetera, to enhance animal production, particularly in the dairy, pig, poultry and ostrich sectors.

2 Mannanoligosaccharides for animal nutrition

Although the study of animal genetics continues to offer advances in animal production, feed options are becoming more restricted by the removal of animal proteins from animal diets worldwide. The banning of some feed additives in Europe, and the increasingly demanding health restrictions on meat products, has made the task of advancing animal production more and more difficult. Mannanoligosaccharides (MOS) represents a new era in biological gut microfloral modifiers. They are derived from the cell wall of the yeast, *Saccharomyces cerevisiae.*

The evolution

Investigations into the effectiveness of yeast culture in monogastric diets has led to the identification of yeast cell wall mannanoligosaccharides as a primary factor in many of the responses. The fruit represents the yeast cell and the peel represents the yeast cell wall. This is separated into the inner (white rind) and outer cell wall (orange-coloured skin). The outer cell wall consists of glucomannoproteins, of which mannanoligosaccharides constitute a major part.

Mannanoligosaccharides are incorporated in the diets of many different species. The standard inclusion rates are 0.5 to 1 kg/tonne (1–2 lb/ton) for turkeys; for broilers, 1 to 2 kg/tonne (2–4 lb/ton) for breeding poultry stock, 0.5 to 1 kg/tonne (1–2 lb/ton) for layers, 2 to 3 kg/tonne (4–8 1b/ton) for piglets, 0.5 to 1kg/tonne (1–2 lb/ton) for growing and finishing pigs, 2 to 4 g/head/day (0.07–0.14 oz/head/day) for calves and 1–2 kg/tonne (2–4 lb/tonne) for rabbits. These modifiers are vital in the livestock industry because sub-clinical infections and challenges continue to prevent animals from reaching their maximum genetic potential. As a result, reliance has been placed upon feed additive antibiotic growth promoters.

A wide range of viable alternatives are being evaluated on the basis of their heat stability, effectiveness at controlling specific and non-specific challenges, and their potential to enhance performance. Oligosaccharides, particularly MOS, are the most promising of those studied to date.

These oligosaccharides include a wide range of molecules, which are natural constituents of plants and micro-organisms, such as yeast. Oligosaccharides are sugar complexes, which contain a small number of similar glucose, fructose, and mannose monosaccharide units, arranged in either a linear or branched structure.

3 MOS Effects on livestock

Mannanoligosaccharides has positive effects on the weight gain and feed efficiency of turkeys, though not at higher inclusion levels. In broiler production, MOS command very high growth rates (about 5% heavier) and improve on feed conversion and very low mortalities. MOS can command over 1 kg heavier pigs at weaning than those not fed.

Bacterial attachment

Feeding yeast-derived MOS is a cost-effective form of mannanoligosaccharides. MOS can defeat the first phase of infection in a similar manner to mannose without being utilised or fermented. Some mannan-sensitive bacteria that have been identified over the years include *E.coli* strains 15R, K99, 078:H12, 06:K13H31, NCSU 1G1-5, NCSU 1G8-11, NCSU 1G3-3 and NCSU 1G8-2. MOS have shown the ability to modulate immune function in a wide range of species.

Public concern about antibiotic resistance in human pathogens continues to spread. Consumer activists and government regulatory officials are blaming the use of antibiotics as animal performance enhancers as a major cause of the resistance problem. For this reason, a growing number of countries have banned the inclusion of sub-therapeutic concentrations of antibiotics in livestock and poultry feeds.

Fortunately, mannan oligosaccharides provide livestock and poultry producers with an alternative option to the exclusive use of feed-grade antibiotics. Mannanoligosaccharides prevent bacterial infections via mechanisms that are different from those of antibiotics, therefore, circumvent the pathogen's ability to develop resistance. According to numerous laboratory and field trials, mannanoligosaccharides can improve performance, either used on their own or in combination with a growth promoter.

4 Anti-coccidiosis

Anticoccidials containing the unique polyether ionophore lasalocid offer excellent coccidiosis control and optimal bird performance. These are recommended for the prevention of coccidiosis in broilers, replacement pullets and turkeys. Dosages of such anticoccidials range from 75 to 125 ppm of lasalocid.

Lasalocid results in a sustained activity against all coccidial species in broilers, turkeys and replacement pullets. These results persist even where other ionophores seem to have become less effective. Lasalocid exhibits good control of all species occurring in the broiler, especially against caecal coccidiosis.

Lasalocid is lethal to coccidia early in their life cycle. This is before they cause excessive damage. Its use results in significant growth response and improved feed conversion, even under high challenge conditions. A very important attribute of lasalocid is that there is no cross-resistance between lasalocid and other ionophores and, therefore, is very suitable in rotation programmes.

Lasalocid has a wide margin of safety. This allows for a wide dosage range. Unlike other products, lasalocid does not increase the requirements for certain nutrients such as amino acids, hence permits optimal feed and water intake, thereby optimising bird performance, particularly in warm weather.

5 Enzyme technologies

New enzyme technologies have been developed allowing maximization of the full potential of pig and poultry diets. This is through improving digestibility of the vegetable protein sources.

The impact of enzymes on soya meal

Soya bean meal is the common denominator in pig and poultry diets. The energy value of soya bean meal is relatively low. This is due to the presence of different anti-nutritional factors. The nutritional value of soya bean meal can be improved with the use of enzymes by improving amino acid digestibility. Digestibility of cystine, threonine, lysine and methionine can be increased by over 5% when 1 kg of an enzymatic product is mixed per tonne of feed.

Enzymes used to enhance feed utilization normally have peculiar features and benefits, which have been summarised in table 16.1.

Table 16.1 Characteristics of enzymes used in animal nutrition

Features	*Benefits*
Improved energy utilization of vegetable protein	Improved livestock performance Reduced feed costs (e.g. fat and space)
Improved nitrogen utilization of vegetable proteins	Improved livestock performance Reduced feed costs (e.g. amino acids) Reduced faecal nitrogen loss
Improved raw material utilization	Higher use of least cost materials e.g. beans, oilseed rape Breakdown of anti-nutritive factors Better litter quality in pigs.
Reduced faecal nitrogen	Reduced downgrading of carcass Lower ammonia levels Improved environment
Low inclusion in powder form	Ease of application into feeds either directly or through a premix

Nutritionists can decide, in view of the foregoing enzyme attributes, to either improve growth and performance by addition of the enzyme powder on top of existing diet, or to reformulate the diet to take account of the additional energy and protein digestibility with the enzyme.

Enzymes and micro-encapsulated bacteria

Enzymes and micro-encapsulated bacteria can be used for natural silage. These are the most advanced silage inoculants that are now available on the market for use in livestock production. The result of using these biotechnological products is to ensure perfect silage making every time.

The main reason why enzymes are useful is that they ensure that the bacteria are provided with sufficient sugars and nutrients for rapid growth, while the power of micro-encapsulated bacteria lies in that this product ensures that the bacteria are alive and fresh when applied – just as when they were produced. Of course, the danger with encapsulated bacteria is the flourishing of opportunistic bacteria, like clostridia, that make bad silage. These must be prevented because they lead to loss of valuable energy and protein due to poor fermentation.

The lactic acid bacteria encapsulates are the type of bacteria that are responsible for the making of good silage. The quicker they can be made to grow the better. Ensuring that the right bacteria are present in large numbers in the silage means that the growth of bacteria such as clostridia are prevented. The application of encapsulated bacteria is recommended at 5g/tonne for maize and sorghum silages. The application rate is increased to 10g/tonne when silage is made from grasses in general.

The procedure of application is guided by the following:

1. Dilute contents of 1 sachet in 10 litres of cold water, and stir well,
2. Pour the solution into the container and add water up to
   i) 100 litres for maize and sorghum silage or
   ii) 50 litres for other silages,
3. Application is 2 litres of solution per tonne forage,
4. Apply during chopping or when loading into the silo or silage pit,
5. Applicator must be thoroughly cleaned with clean water,
6. Never mix with other biological or chemical agents,
7. Mix only enough, as required as the product will only remain active for 48 hours after dissolving,
8. Use continuously during cutting or ensiling processes,
9. Never use warm water because it destroys the micro-organisms.

These biotechnological products work well where general silage making management is good. This situation then guarantees that their use will improve daily weight gains, improve digestibility of the silage, better utilization of energy and protein, and improved milk production. All these desirable results stem from improved fermentation of the silage, reduced fermentation losses, preservation of more nutrients, inhibition of clostridia activity and reduced acetic acid formation. Experience has also shown that the resultant silage is tastier and hence leads to higher intake and improved roughage digestion.

6 The power of seeding feed with Yea Sacc yeast

Yea Sacc1026 is composed of yeast cells from the unique yeast strain *Saccharomyces cerevisae*1026 strain. This is grown in batch cultures on a medium of maize, molasses, malt and trace minerals. The result is a concentrated mixture of yeast cells and cell metabolites produced during fermentation. This mixture is then carefully dried so as to maintain the viability of the yeast cells.

Yeast cells have the ability to alter the bacterial population that occurs naturally in the digestive tract of animals. These bacteria are usually responsible for the fermentation of otherwise indigestible feed in the digestive tract. Specific strains of yeast have unique effects on the bacterial population.

Actively metabolising yeast cells produce cell metabolites that stimulate the growth of certain bacteria. *Saccharomyces cerevisae*^{1026} has the ability to stimulate the growth of cellulolytic bacteria (fibre digesting) in the rumen and lactic acid utilizing bacteria.

The following are the biochemical functions of yeasts. They:

1. Decrease lactate production;
2. Favour the flourishing of cellulolytic and lactic acid utilizing bacteria, and increases rate of cellulolysis in the rumen;
3. Lower methane production in the rumen by altering the methanogenesis process itself;

4. Improve rumen pH stability;
5. Change volatile fatty acid proportions producing more of acetate, propionate and butyrate in decreasing order;
6. Increase feed intake and improve animal productivity.

Yeast cells are used mainly in ruminant diets. Because yeast has the ability to stimulate the growth of fibre digesting and lactic acid utilizing bacteria, yeast cells stabilise and optimise rumen fermentation. This is particularly the case in ruminants that are fed rations that contain high proportions of grain (e.g. early lactation diets fed to dairy cows). High grain (energy) diets tend to cause a drop in rumen pH due to the production of lactic acid, and this has a negative effect on fibre digestion. By stimulating the growth of cellulolytic and lactic acid utilizing bacteria, yeast is able to stimulate feed intake, increase microbial output from the rumen and increase milk production. Responses do vary with differing diets, but generally, an increase of 1–1.5 kg of milk/day/cow can be achieved. Milk butterfat can also be increased, especially in situations where fibre digestion is suppressed.

With monogastric animals, the situation is slightly different but yeast cells still produce desirable effects. Monogastric animals such as pigs, horses and ostriches have active microbial populations in their hindgut (caeca), which digests a significant amount of fibrous feed. Caecal fermentation is important because the fibre digestion which takes place there is of great importance to these animals. By stimulating the bacteria, just as in the rumen, feed intake and utilization is improved. The energy contributions from hindgut fermentation in monogastric animals are small, but the minerals, especially calcium and phosphorus that are released from the feed by this fermentation and then absorbed by the animal, could be very important.

The recommended application levels are 10 g/animal /day for ruminants and 1 kg–2 kg per tonne of feed in monogastric animals, especially in times when intake is low. In pigs, this can prove to be very beneficial in starter diets or shortly before farrowing.

7 Lactic acid bacteria as biological silage inoculants

Crops and pastures only grow during certain seasons of the year and soon after maturing their nutritive value starts declining. Making silage before this deterioration offers the farmer the greatest opportunity to harvest these forages when their nutritional value is still high. Ensiling preserves the feed for later use in the dry periods of the year when there is no grazing to talk about. This is particularly important in most tropical environments.

Ensiling is achieved through a fermentation process. During fermentation, soluble carbohydrates are converted to lactic acid, which drops the pH of the silage to a level where any further bacterial activity is inhibited. The forage is, therefore, preserved by the lactic acid. This conversion of soluble carbohydrates to lactic acid is done by lactic acid bacteria under anaerobic conditions. As a result, inoculants in the form of lactic acid bacteria make the lactobacillus dominant and in turn crowd out clostridia, coliforms and yeasts.

Examples of lactic acid bacteria are *Streptococcus faecium*, *Pediococcus acidilactici* and *Lactobacillus plantarum*. The first two are very active and hence most preferred because of their ability to begin fermentation quickly and under aerobic conditions. They will, therefore, begin producing lactic acid right at the beginning of the ensiling process. This brings the pH down into a range where the growth of *Lactobacillus plantarum* is stimulated. *L. plantarum* is the primary lactic acid producer and is, therefore, able to cause a rapid drop in the pH to between 4.4 and 5.0 in the case of small grain and legumes and 3.8 to 4.2 for grass and maize silage. This rapid fall in pH is vital since the undesirable bacteria that spoil the silage cannot survive at such a low pH.

8 Use of deodorases in livestock feeds

As the name implies, deodorases are enzymes that have anti-odour properties. The enzymes are capable of binding ammonia and many acidic compounds such as hydrogen sulphide, volatile fatty acids and other noxious gases that create disagreeable odours in feeds. They can be used in very intensive systems where large amounts of ammonia and other gases are a natural consequence of keeping animals in confinement. Odours due to ammonia easily occur because intensive livestock systems use high protein diets that are easily degraded or decomposed into ammonia gas. If ammonia levels in the air exceed 6 ppm, eye and respiratory membrane irritation begin. Animal performance starts to decline at 11 ppm. At 50 ppm severe reductions in animal performance and health become evident. At this high level, there is also the possibility of pneumonia when levels reach 100 ppm; then the animals are seen sneezing, salivating, and get irritated in the mucosal surface.

The crucial level for commercial farmers is 11 ppm. At this level, ammonia begins to have a negative effect on animal performance. Experience has shown that in most cases ammonia levels are higher than expected and farmers become accustomed to it because of daily exposure.

Some deodorases can lower ammonia levels by up to 50%, over a 2–3 week period. This has, in turn, translated into improvements in animal health and performance in all types of poultry housing, as well as in all stages of pig production.

In ruminants, deodorases assist in the efficient use of ammonia for protein synthesis by rumen bacteria, particularly with diets that include high levels of NPN (non-protein nitrogen).

9 A special look at biotechnology in poultry production

Biotechnology relates to the application of biology to production. In the poultry industry, biotechnology can be used as a tool to meet the challenges of increasing productivity and efficiency. Although its adoption in terms of poultry nutrition has been slow, biotechnology has a lot of potential in developing countries as it offers many advantages, which include the following benefits:

- Maximising efficiency of conventional raw materials used by poultry;
- Improving utilization of non-traditional raw materials in poultry diets;
- Reducing air and land pollution; and
- Increasing meat and egg quality and consumer perception.

Odour control

The quality of air in poultry houses is very important, especially during cold season brooding. This is because in winter the producer is faced with the need to conserve warmth by reducing ventilation, thereby letting the birds live in high ammonia level conditions, or increasing ventilation rate, which drops the temperature in the house. High ammonia levels (>25 ppm) decrease broiler growth rate and feed conversion efficiency, and cause breast blisters which can lead to downgrading of carcasses and increased condemnation.

A biological approach to get around this problem is available through products which bind ammonia and other noxious gases released from poultry waste, as alluded to in section 8. These products can be incorporated in poultry rations. In addition, these products also act on ascites (water belly) in broilers, via their effect on blood oxygen capacity, which helps to reduce depressed performance and high mortality associated with this disease.

Acidifiers and probiotics

Most bacteria of a pathogenic nature, such as salmonella and coliforms, will survive at high pH, while beneficial bacteria thrive at low pH. When fed to poultry, mostly via drinking water, acidifiers lower the pH of the gastrointestinal tract and this will in turn favour its colonization by non-pathogenic bacteria such as lactobacillus. When acidifiers contain a live yeast culture they are called probiotics because the yeast comprise beneficial micro-organisms – 'the probiotic effect'. In a mixture with acidifiers and electrolytes, probiotics occupy the gap in the microflora after medication, prohibit pathogenic bacteria from multiplying and alleviate the effect of stress, respectively.

Feed toxins and contaminants

As has been mentioned in chapter 12, mycotoxins are toxic chemicals produced by moulds (fungi). Mycotoxin contamination occurs the world over, affecting crops in the field, in storage, and during feed manufacturing, delivery, and feed mixing on the farm. Poultry are affected by mycotoxins in different ways, which cause production losses.

The levels of toxins detected are generally near the lower limits of detection. Therefore, the use of toxin binders in poultry diets seems to be of significant value as an insurance against some of the possible adverse effects of mycotoxins on performance. Commercial mycotoxin absorbants are available for both feed millers and on-farm mixers. Ideally, a binder should be compatible with the feed, bind a broad spectrum of toxins at low inclusion (0.5–1.0 kg per tonne of feed), and should not tie up with other nutrients, especially minerals and vitamins.

Enzymes

Enzymes are biological catalysts which increase both efficiency and rate of digestion and metabolism. In poultry production, feed constitutes up 80% of total costs of production, and, coupled with the rising cost of feed ingredients, enzymes have a big potential for improving dry matter digestibility, resulting in better utilization of feed by the birds. Use of exogenous enzymes also improves energy metabolism, reduces digesta viscosity and minimises the detrimental effects of anti-nutritional factors present in both conventional and non-conventional feed ingredients. Litter quality is also positively affected as it becomes drier.

A variety of enzymes such as proteases, carbohydrases (including fibre digesting enzymes), lipases and phytases are available on the market. These can be mixed and matched to form a cocktail to fit any particular diet, but their impact is more pronounced in low nutrient density rations.

Other products

Other biotech products are available that improve eggshell strength, yolk quality (pigments), reproductive performance in breeder flocks and overall flock health (chelated minerals, yeasts), and growth in broilers (antibiotics, ionophores).

Biotechnology therefore seems set to play a key role, not only in terms of poultry nutrition, but also in other livestock species and crop husbandry as we strive to improve the range of animal feedstuffs and the efficiency with which animals utilize these feeds.

10 Conclusion

Under the modern day economic conditions, it has become increasingly important to maximise production. Biotechnology has been used successfully in the developed world, to improve margins and profits in incredible ways. Livestock production is no exception. The objective of using biotechnology in animal feed science is to maximize effectiveness and efficiency of nutrient utilization and hence, productivity.

References

Cowan, D. 'Industrial enzyme technology', in *Trends in Biotechnology* 1996, **14**:177–178.

Hay, V.W. and W.M Muir, 'Efficiency and safety of feed additive use of antibacterial drugs in animal production', in *Canine Journal of Animal Science*, 1979, **59**:447–456.

Hodgson, J 'The changing bulk biocatalyst market', in *Bio/Technology,* 1994, **12**: 789–790.

McDonald, P., R.A Edwards, and J.F.D Greenhalgh, *Animal Nutrition*, fifth edition, Harlow Longman, 1995

Mutetwa, L. 'Biotechnology in Poultry Production', in The INC Newsletter, December 2000.

Petterson, D. and P Aman, 'Enzyme supplementation of a poultry diet containing rye and wheat', in *Br J. Nutr*, 1989, **62**:139–149.

Rode, L.M, ed., 'Use of feed enzymes in ruminant nutrition. by Beauchemin', K.A. and L.M Rode, in *Animal Science Research and Development – Meeting Future Challenges,. Proc. Can Soc. Anim. Sci*, Lethbridge, pp.103–130, 1996

Topps, J.H. and J Oliver, 'Animal foods of Central Africa', in *Technical Handbook No. 2, Zimbabwe Agriculture Journal.* 1993

Wallace, R.J, 'Ruminant microbiology, biotechnology, and ruminant nutrition: progress and problems', in *Journal of Animal Science*, 1994, **72**: 2992–3003.

18

STOCK FEED MANUFACTURING

1 Introduction

The objective of this chapter is to provide information on a number of different types of roughage processing techniques and stock feed manufacturing methods. Emphasis will be placed on those technologies perceived to be most important to present day and future agricultural practices. Discussion will centre on how the modifications improve food utilization by livestock (especially ruminants) (see also chapter 15), the potential for further improvement and economic competitiveness in the future.

2 Stock feed processing and manufacturing

Stock feed processing can be defined as the modification of livestock feed ingredients, especially crop seeds and crop residues after harvesting. Manufacturing is the industrial production of any of the following:

- Additives and supplements
- Blocks/cakes/cubes/pellets/flakes
- Complete feeds
- Concentrates
- Medicated feeds

Modification of forage quality should place emphasis on a number of factors. Some of the key ones are:

- Target material

- Purpose: e.g. to increase acceptability of fibrous feeds so that there is an increase in daily intake; this will enhance rate and (or) extent of digestion and consequently improve nutrient availability.

The ultimate aim of modifying feed ingredients should be that the processed material should be competitive nutritionally and economically with conventional feeds, and/or natural feed for a particular type of animal.

3 Types of feed processing

The most important types of feed processing are physical treatments, chemical treatments and biological treatments. Physical treatments include grinding, pelleting, steam treatments and mechanical separation of plant parts.

Grinding and pelleting:

- Decrease particle size so feed is pre-digested for the animal,
- Increase surface area for efficient enzyme action during digestion,
- Increase the bulk density of feed fractions so that animals take in more nutrients per unit time, hence using less energy in feeding.

Milled feeds are normally pelleted before feeding to further increase bulk density, decreased dustiness, improve on ease of handling. The most consistent animal responses to grinding and/or pelleting feed has been marked increases in feed intake, daily weight gain and food conversion efficiency.

Increase in intake has been shown to be proportional to a decrease in particle size up to a critical size of 1 mm. This is more important in ruminants where digestibility will thereafter decrease with grinding, due to reduced fibre digestion in the rumen. This is caused by shorter gastrointestinal tract residence time, hence enzymes have less time to digest the feed.

Steam treatment

This involves steam pressure treatments of forage or crop residues. Steam treatment softens the fibrous material and hence improves the utilization of lignocellulosic materials. The treatment conditions are normally 5–40 kg cm^{-2} for less than 5 minutes. A variation to this process is a steam explosion process. This is where the pressurised feed material is quickly exposed to atmospheric pressure. Steam treatment modifies feed by a physical disruption of cell wall structure and through a hydrolytic action on lignocellulosic complexes. Acetyl esters of hemicellulose are cleaved, thus releasing a considerable amount of acetic acid. This lowers the pH of the substrate and assists in hydrolysis.

Mechanical separation of plants

The aim of this method is to achieve separation of discrete fractions of differing quality for specific uses, e.g. leaves from stems. This involves use of controlled velocity air currents in closed systems to separate light (leaves) from heavy (stem) particles.

Chemical treatments

Chemical treatments modify feeds through a number of reactions like:

- hydrolytic action
- ammoniation and
- oxidative activity

Chemical treatments are more beneficial with mature, lignified substances. Material from monocots has been found to be more responsive to chemical treatment than material from dicots.

Hydrolytic action

The common hydrolytic agents are sodium hydroxide and ammonium hydroxide. Alkali treatment causes partial solubilisation of hemicellulose, lignin, silica and hydrolysis of uronic and acetic acid esters. It also results in a disruption of intermolecular hydrogen bonding in cellulose and therefore increases hydration of treated forage materials. Hydration is necessary for effective microbial colonization of fibrous materials and hence results in efficient degradation of the materials in the rumen.

Ammoniation

Ammoniation can be done with urea fertilizer or other sources of ammonia like anhydrous ammonia and ammonium hydroxide. The mode of action is similar to the hydrolysis of fibre discussed above, but ammoniation has the added advantage that it improves the feed quality by boosting the nitrogen content, and hence the crude protein level of the feed, reducing the need to provide supplemental nitrogen through expensive protein concentrates. Factors to consider in ammoniation are:

- Amount of ammonia: requires 20–30 kgt^{-1} of feed material to be treated.
- Water content: requires 293kgt^{-1} of material to be treated. A solution is made and sprinkled on feed material, with occasional compressing in silage pits or plastic bags, just as it is done in silage making.
- Temperature: the higher the temperature the better. Ammoniation is better done during the hot dry seasons, especially in tropical environments.
- Treatment time: the treated material is incubated for three to four weeks in the silage bags or pits. Urea breakdown releases ammonia. Ammonia dissolves in water and forms a strong alkali ammonium hydroxide. Ammonium hydroxide reduces the fibrosity of the feed (i.e. soften it so that animals will not waste energy chewing the material for a long time). Nitrogen content (hence, crude protein) is increased.
- Type and quality of substrate: it is more advantageous to ammoniate very fibrous straws than good quality hay.

For smallholder farmers, the ammoniation process is simplified using 4% urea solution as 1 bag urea fertilizer (about 50 kg) dissolved in 1 drum of water (about 250 litres) and 1 tonne of hay or crop residues.

When the ammoniated material is ready, it will be soft and slightly wet with a characteristic flavour. The pit or bags should be opened for at least two days before feeding the ammoniated stuff to the animals. This is to allow excess ammonia gas to escape so that chances of urea poisoning are eliminated. Environmental problems (pollution) associated with ammoniation are minimal compared to sodium hydroxide. These are limited to the disposal of the plastic covering the stack, than ammonia itself.

Oxidative treatments

Chief oxidation agents commonly used are ozone, sulphur dioxide, chlorite, peracetic acid, hydrogen peroxide and permanganate. These oxidants actively attack and degrade a major proportion of cell wall lignin. Ozone and hydrogen peroxide randomly cleave glycosidic linkages of cell wall polysaccharides. Sulphur dioxide causes polysaccharide solubilisation. The net effect of oxidants is reduction of lignin content and increase in soluble carbohydrate concentration in the modified feed. Mixtures of hydrolytic and oxidants is a promising treatment of the future.

Microbial and enzymatic treatments

Microbial and enzymatic treatment, are novel and highly favoured, mainly due to their advantages over chemical treatments. These advantage are that fewer or no chemicals are used (this saves the environment) and that lower energy inputs are required. The disadvantages are that it takes longer to treat and, since there is active degradation of the material by the microbes or enzymes, there is inevitable dry matter loss. There are also constraints, for example, in finding microbes that can degrade lignin complexes and leave out cellulose. Research with white rot fungi has shown some promise in that area. It is important to select fungi species, inoculum size, length of incubation and a suitable environment so that the treatment objectives are achieved fast enough, and is as inexpensive as is possible. The most useful enzyme in fibrous material (if it can be produced cost-effectively in large quantities) is cellulase. It will be a tremendous biotechnological advancement if lignase could be produced industrially. It, however, remains a priority area of research and is a vital enzyme of the future.

4 Stock feed manufacturing

Manufactured stock feeds come in various forms and types. The most common ones are additives, supplements, blocks, cubes, cakes, pellets, flakes, complete feeds, concentrates, premixes and medicated feeds.

Additives and supplements are single (or a combination of) ingredients which are included in the basic feed mixture to fulfil a particular need. Good examples are urea, lysine, methionine, mono calcium phosphate or dicalcium phosphate. These are normally used in feed mixtures in minute quantities. In most cases they are fed free choice, undiluted or carefully mixed with the ration at point of feeding.

Blocks, cakes, cubes, pellets, and flakes are physical forms of feed presentation. Blocks are feeds compressed into a solid mass cohesive enough to hold its form. They weigh anything more than 1 kg but mostly produced in 15–30 kg forms. Cake is the mass resulting from pressing of oilseeds, meat or fish in order to remove oils, fats or other liquids. Cubes and pellets are produced by compacting and forcing the feed through openings that shape them into cubes and pellets. Flakes are ingredients rolled or cut into flat pieces, with or without prior steam conditioning.

Complete feeds are animal specific diets manufactured (or simply blended) to supply adequate amounts of all the necessary nutrients. By a specific formulae, the feed is meant to be fed as the sole ration and is capable of maintaining life and, in addition, support production without any additional substances being consumed, except for water.

Concentrates are feed ingredients known to be concentrated in a critical nutrient. They are mainly used to improve the energy or protein value of the whole diet. As such, we have protein concentrates like fish meal, meat and bone meal, blood meal, soya bean meal, cotton seed cake, et cetera. We also have energy concentrates like maize meal and molasses.

Medicated feeds are manufactured feeds that contain drug ingredients meant to cure, initiate, treat, or prevent diseases of animals. Others are meant to affect the structure or function of the body of the animals. Note that antibiotics included in a feed for growth promotion and to improve feed utilization efficiency are drug additives, but the feeds containing these are medicated feeds. Good examples of medicated feeds are layers concentrate with Lavadex to check on internal parasites, and layers concentrate with coccidiostat to check on coccidiosis in poultry.

5 Reactivity of feed components during manufacture

During stock feed manufacturing, chemical components in the feed (e.g. carbohydrates and proteins) take part in chemical reactions which must be controlled through online quality control. This is necessary so that the feed is not under processed or over processed.

Carbohydrates, for example, undergo Maillard reactions, caramelisation, solubilisation and resistant starch formation. Proteins undergo Maillard reactions, LAL/LAN formation (see the section overleaf), deamination, D-amino acid formation, iso-peptide formation and denaturation. Lipids undergo auto-oxidation, thermal degradation, cis and trans isomerisation, polymerisation and formation of Maillard reactants.

Why are these reactions important? They are important because they:

1. Produce chemical products that impart flavour and aroma to the feed,
2. Improve palatability of the feed,
3. Destroy anti-nutritional factors,
4. Improve quality, e.g. crude protein content through ammoniation.

Caramelisation and Maillard reactions

Reducing sugars take part in caramelisation and Maillard reactions during moist heat treatments. This has a profound effect on the flavour of feed and results in loss of nutritional value if not controlled.

Caramelisation is a group of reactions that produce caramels, the chemicals that impart colour and aroma to the processed feed. In the human feed industry, caramels are important in the processing of chocolates and sweets or candies.

With starch granules, moist heat treatments causes them to swell or gelatinise. In certain conditions, resistant starch is formed and this is not digestible in monogastric animals.

Amino acids/peptides and proteins containing lysine take part in Maillard reactions. Heat treatment in alkali conditions cause cross-linking of amino acids in proteins. When the cross-linked amino acids are hydrolysed, LAL (Lysinoalanine) and LAN (Lanthionine) are formed. LAL and LAN are resistant to digestion such that this reaction is undesirable. Other unwanted reactions are the formation of non-nutritive D-amino acids from the L-configuration. Semi to dry heating at pH = 7 causes cross-linking which results in iso-peptides.

Maillard reaction involves the reactions between reducing sugars and free amino groups from amino acids and oxidation of vitamin C, forming reducing compounds and oxidation of polyphenols. They are largely undesirable and must be controlled, but essential in the formation of characteristic colour and characteristic aroma/flavour in baking, cooking and roasting. Over Maillardation results in the loss of amino acids, reduction of digestibility through cross-linking of proteins, reduction of bio-availability of iron and zinc and formation of toxic compounds.

Protein denaturation and oxidation

Denaturation coagulates and inactivate bioactive proteins like enzymes, lectins, trypsine inhibitors. Lipids are very prone to oxidation especially when feed contains unsaturated fatty acids. Heat treatment speeds up chemical oxidation, and prevents enzymatic oxidation by rendering lipoxygenase inactive. Products formed in chemical oxidation can take part in Maillard reactions and results in off flavours.

Oxidised lipids can react with proteins and reduce their digestibility. Increased viscosity occurs due to isomerisation of double bonds from the cis to the trans-configuration, dimerisation, and polymerisation.

Protein denaturation occurs during thermo-mechanical treatments. This is a change in the conformation of a protein that does not involve the breaking of bonds. In most cases, proteins undergo structural unfolding, aggregation due to moist heat and shear effects.

The three dimensional structure is destabilised as a result of heat and shear by breaking hydrophobic and electrostatic interactions. Denaturation is accompanied by the uncovering of hydrophobic groups and this results in a decreased solubility of the proteins in aqueous solutions. If the mechanical energy is very high, denaturation is followed by hydrolysis of peptide bonds, modification of amino side chains, formation of new covalent iso-peptide cross-links.

Reactivity of minor components

Minor feed components refers to those components found in minute quantities and is not a measure of importance. Minerals act as catalysts and, due to processing their bioavailability, may greatly be changed. Some are involved in the Maillard reactions and lipid oxidation.

The stability of vitamins in hydrothermal and thermo-mechanical processed feed depends on their individual chemical properties such that recovery rates vary from 100% to less than 20%. Enzymes used in animal feeds are denatured under moist heat and shear conditions, especially when temperatures used are above 75°C. As a result, it is generally recommended to supply supplements containing these compounds normally in liquid form after processing.

6 Legislation in the stock feed industry

The feed manufacturer is bound by various pieces of legislation which do not vary from country to country that much. The whole idea is to protect animal life and, subsequently, human life. Most importantly, feed contamination should be avoided at all cost. The following contaminants should be avoided and their absence or permitted levels assured by the manufacture:

- Level of aflatoxin B1 should be very low or non existent;
- maximum permitted content of arsenic is 20mg/kg and for cadmium it is 10mg/kg,
- Ammonium sulphate should not exceed 0.5 % incorporation for young ruminants;
- DL – methionine should not exceed 3%;
- L-Lysine and DL – methionine should not exceed 3 %,
- Microbes (single cell protein) should be identified by the strain according to a recognised international code;
- For enzyme inclusion, identification number and activity units should be shown on package clearly;
- For pesticides, the maximum residual level in crop and feed itself should be noted
- The commonest pesticide found in cereals is pirimiphos-methyl and its maximum residual level permitted is 5000 ppb,
- There is also a dietetic feedstuff directive stipulating that the total feed should:
  - reduce risk of milk fever,
  - reduce risk of tetany (hypomagnesemia),

- reduce risk of acidosis,
- reduce risk of urinary calcium,
- reduce risk of constipation,
- reduce risk of fatty liver syndrome,
- reduce risk of digestive disorder of the large intestines.

7 Quality control

Quality control is critical in order that the feed manufacturer adheres to certain standards. It is necessary to monitor thermo-mechanical treatment in particular so that the food is not over or under processed. Desirable animal feed products result from the control of temperature, residence time of feed in a particular section of plant or equipment, moisture content and shear forces. So assuring quality is imperative and this is done through routine or spot on analysis to monitor chemical, physical changes, destruction of anti-nutritional factors like anti-trypsine inhibitors, optimum Maillard and/or denaturation. The manufactured feed product should not be over or under cooked.

Quality control methodologies

In vitro (in glass) digestibility predicts in vivo (in the animal) digestibility values. A rapid multi-enzyme technique that involves measuring pH drop after adding proteolytic enzymes to samples containing the processed proteins is done and this is then related to in vivo (in the animal) digestibility. A variation to this method is when pH is kept constant during enzymatic incubation. The amount of NAOH needed to neutralise the production of hydrogen ions is used as an indication of in vivo digestibility.

Urease activity is a test phenomenon that has been used for years to monitor heat treatment of soya products. Normally heat treatment of soya bean is meant to destroy trypsine inhibitors. Reduction in urease activity, i.e. less ammonia production, hence higher pH in ammonia solution, occurs due to urease destruction. The principle behind the method is that destruction of urease occurs at the same time as trypsine inhibitor inactivation. It is measured as a rise in pH. Urease activated is a satisfactory measure of heat treatment of soya products but can not help to predict cross linkages.

Protein dispersibility index (PDI) is useful in this regard. The principle is that protein denaturation is accompanied by a decrease in protein solubility due to cross linkages. So the PDI basically measures the proportion of protein which is dispersible in water. PDI of raw materials is 80% and that of treated materials is around 20–40 %.

Nitrogen solubility index (NSI) is another method used to predict cross linkages. The method involves extracting proteins in alkali solutions using 0.042M KOH. NSI values greater than 85% indicate under processing and NSI less than 75% is unsuitable. Feed with such low NSI tend to depress growth of livestock. NSI is a better index than PDI, while PDI is better than in vitro digestibility.

Solubilisation measurements are sometimes done to monitor protein-protein interactions. Examples of these interactions are hydrophobic interactions, electrostatic interactions, hydrogen bonding, dipole-dipole interactions, van der Waals forces. During thermo-mechanical treatments, these are broken. So the method is based on solubilisation of denatured molecules using reagents with well known mode of action like urea and sodium sulphite.

8 Conclusion

Stock feed manufacturing should be treated with due seriousness and care, just as food processing is for human beings. So, care has to be taken in ensuring that the desired reactions occur during processing. These must be controlled so that the correct processing levels are achieved. Quality control is just as critical as in the human food industry.

References

Bellis, D.B. and R.B Brooks, 'Farm Processing of soya beans for pigs', in *Rhodesian Agricultural Journal*, **1975, 71**:99.

Maeng, W.J, D.S Kim, and S.C Hwang, 'A study on the physicochemical treatment of straw in beef and sheep', *Proceedings of the eighth World Conference on Animal Production,* 1998, Seoul National University, June 28-July, pp. 32–33.

Volk, S. and O Khariv, 'The effect of heat treated conola meal supplementation in diets for bull calves on growth rate and blood metabolite level', *Proceedings of the eighth World Conference on Animal Production*, 1998, pp.18–19.

18

ESTABLISHMENT OF STOCKFEED MANUFACTURING OPERATIONS: A CASE STUDY

1 Introduction

This chapter is going to be tackled as a case situation, where the assumption is made that the organization that is setting up stock feed operations is a parastatal that deals with grains (cereals and oil seeds) in a developing country. The section on equipment inventory will therefore assume that the operations are being set up in Zimbabwe, a tropical Southern African country that is developing but has great potential in stock feed production.

Most companies that trade in grains are well conditioned by abundant local raw materials for stock feed production and a large consumption market; at the same time they are beneficiaries of development investments incentives by virtue of their close relationship with governments. If there are few big players in the stock feed industry of any country, and maybe with a hoard of other small-time players, a grain trading company can easily enter into this market by virtue of the above and other unique characteristics provided the correct things are done right from the beginning. The main stock feed products in demand the world over are protein concentrates, energy concentrates, formulated concentrates, complete feeds, blocks, cakes, cubes, pellets, flakes, crumbs, additives, medicated feeds, minerals, vitamins, hays and straws.

The objectives of a new player should be to establish and operate a vibrant livestock feed manufacturing division or subsidiary company that will produce highly nutritive feeds to complement or to replace the role played by monopolies.

The justification for new entrants in such an environment will be that as more players join in, they stabilize or force other operators to become more client centred, more efficient and hence charge lower prices for the benefit of the farmers. Many grain traders have unique potential in stock feed manufacturing since they control the two most important ingredients in stock feed manufacturing, which are maize and soya beans (plus their residues). Most of these grain traders are government parastatals or companies and as such normally have depots everywhere throughout the country. As a result, they are well positioned to be a vehicle for the development of livestock production among the traditional smallholder farmers and the various classes of medium to large scale farmers.

2 Project description

Livestock, like any other animals, require well formulated rations that provide the needed ingredients which are commensurate with the kind of production levels the farmer expects. A feed manufacturer needs, therefore, to set up an operation with the right kind of staff, suitable equipment, relevant database of raw materials, and institutional collaboration like relevant ministries and national standards associations. The set-up should be backed by close adherence to the relevant legislation and international quality standards. The process starts with diligent sourcing of resources, processing into specific feeds, and should not end with marketing, as customer feedback is vital for the continued existence of the stock feed operations. In a nutshell, the foregoing describes the kind of project that must be put in place when setting up a stock feed operation.

3 Inputs needed

Quality raw materials, efficient and effective equipment, adequate capital and trained staff are required. To lead this project, an experienced animal nutritionist is needed at head office, backed by engineers, analytical chemists, marketers and trained animal scientists as specialist technical executives. The physical facilities required include access to laboratories, raw material storage facilities, mills, processors, feed coolers, driers, dust blowers (cleaners), weighing scales, blenders (mixers), bagging equipment, vehicles (fork lift and haulage trucks) and end product storage facilities

4 Organisation and management:

An organizational structure that will allow speedy decision making and continuous monitoring and evaluation in addition to instituting total quality management is imperative. A chief nutritionist assisted by animal scientists and other key personnel like engineers and chemists will be ideal. All the workers, from cleaners right up to the chief should be well informed and trained in quality management, in order that total quality management is achieved in the new company.

5 Costs

A brilliant project cannot take off if costs are not calculated properly. This should be done in as complete a fashion as is possible and should take into account inflationary pressures so that the money set aside will be sufficient to set up the project. Costings should be done on equipment, laboratory analysis, consultancy fees, staff recruitment and other overheads.

6 SWOT analysis

Strengths, weaknesses, opportunities and threats (SWOT) must be identified and analysed carefully. This will allow the new stock feed company to position itself strategically so as to make an instant but lasting impact on the livestock feed market. In addition to SWOT analysis, risk management is an important aspect and must receive due attention. However, stock feed operations are modest endeavours with technology that is well established. Therefore, the risk of not meeting the objectives is practically negligible. Environmental impact assessment is another important area requiring attention. It will be suicidal to set up operations that are perceived to cause environmental degradation and pollution. This might easily provoke civil protests.

Implementation scheduling is vital as it will be used as the basis for monitoring and evaluation. Realistic time frames per each activity will unfold into a successful project. This will also allow harmonious allocation of resources, especially in those activities that must take place simultaneously or overlap at some point.

7 Stock feed manufacturing process

The schematic diagram overleaf shows all the important processes that are used in manufacturing stock feeds. An equipment inventory is included. This scheme of feed processing events should be studied in conjunction with the details given in chapter 17.

Stock feed manufacturing process

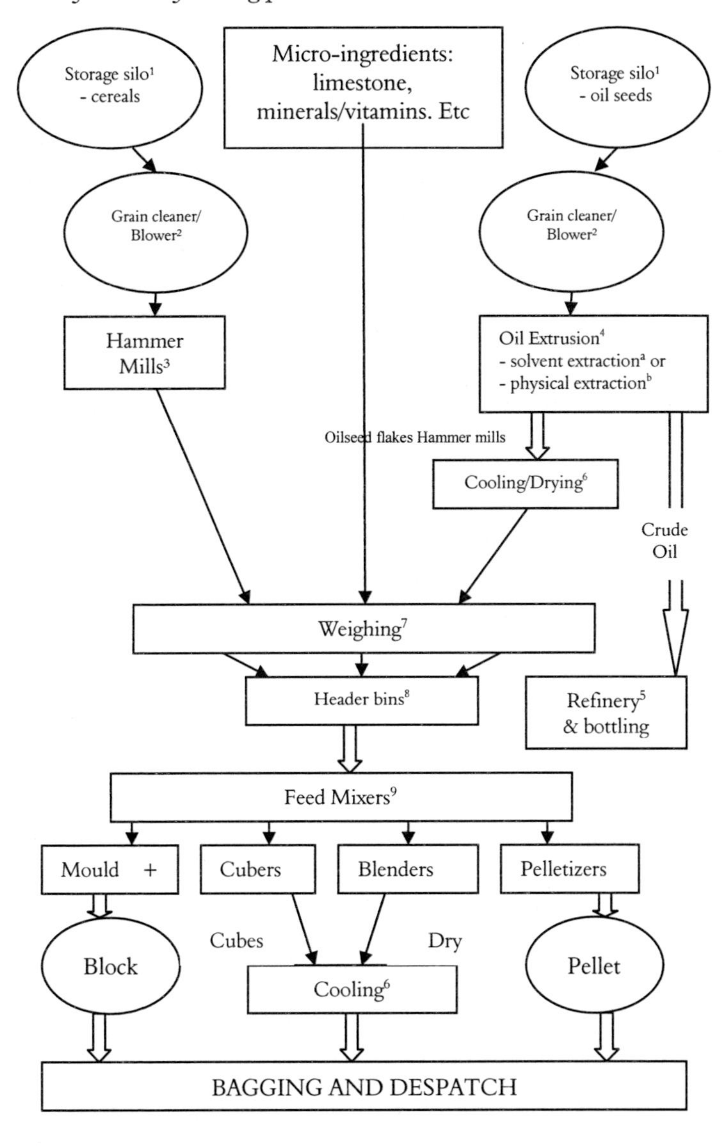

8 Equipment inventory

This section, as alluded to in the introduction, assumes that equipment sourcing is done in Zimbabwe and as such the essential pieces of equipment will be sourced first, from within Zimbabwe if available. If the equipment is unavailable locally, then focus turns on the Southern African region before going international. This is vital for any entrepreneur in a developing country as it is essential to save on cost, foreign currency, and also on time so that implementation scheduling is not derailed.

Storage silos

Maize and oilseeds (soya bean, cotton seed and sunflower) are the major ingredients in the manufacture of stock feed. Supply of these grains is seasonal, hence they have to be stocked up and stored in such a manner that they won't deteriorate in nutritive quality. The storage environment should be kept dry and free from pests. Grain silos are the best storage facilities. In Harare, Zimbabwe, Taylormade (PVT) Ltd[a] manufactures grain silos of capacities ranging from 45 to 200 metric tons.

Grain cleaners/blowers

Foreign matter in grain tends to affect the quality of the finished product, as well as exposing more delicate equipment in the production chain to physical damage. Various grain cleaners are available on the market. Locally, DICHWE Implements[b], Precision Grinders Engineers[c] and Taylormade[a] manufacture grain pre-cleaners that remove dirt from grain through a blowing mechanism. In addition, Taylormade[a] manufactures electric-powered oscillators, which separate grain from impurities by their densities as well as a winnowing mechanism. Magnetic grain cleaners are not available locally. However, Bühler AG[h], a Swiss-based engineering firm with a subsidiary in South Africa, manufactures drum magnets that are used for separating grain from metallic impurities before the grain is fed into other processors. Metal objects damage

[a] See Equipment Supplier Directory

equipment like mills and should be removed from raw materials during the cleaning process. Cleaning of feed ingredients is vital. Animal scientists do not subscribe to the myth that animals will eat anything, including dirt. In actual fact, animals have distaste for filthy and dusty feed. In addition, this can cause respiratory diseases, among many other health problems in animals. What you feed your animal is what you eat at the end of the day.

Hammer mills

In order to ensure uniform mixing with other ingredients, the hay, crop residues, cereal grains and oilseed flakes have to be milled. It is ideal to have separate hammer mills for grinding the cereals and oilseed cakes in order to avoid cross-contamination of the ingredients. In Zimbabwe, engineering firms like Precision Grinders Engineers[c], Taylormade[a], Intermediate Technology[d], Product[e], and Tanroy Engineering[f] manufacture hammer mills of various capacities. The work rate of the hammer mills from these firms varies. One of the highest performers is the Hippo Model 550 from Precision Grinders Engineers[c], which is capable of milling between 900 to 1350 kg of maize per hour. The 20HP Tan-Tan electric hammer mill from Tanroy[f] has the capacity to mill between 600 and 900 kg of maize per hour. The HD10 hammer mill manufactured by Taylormade[a] is capable of milling 500 kg of grain per hour. There is a host of other less performing grinding mills manufactured by these listed firms.

Oil extrusion

The high oil content in oilseeds (soya bean, cotton seed and sunflower) is not desirable in commercial stock feed manufacturing. Instead, oilseed meals are used as ingredients. Oilseed meals are obtained by grinding the cake, chips or flakes after removal of most of the oil from the seeds by a solvent extraction or a mechanical process. Some feed manufacturers now use full fat soya in their formulations. This should, however, be toasted/roasted through an extrusion unit.

The processed full fat soya should then be treated with suitable anti-oxidants to prevent rancidity and hence prolong shelf life. Extrusion destroys anti-nutritional factors.

Solvent extraction

During the solvent extraction process, the seeds are cracked and the hull is removed. The seeds are then moved to tanks where they are soaked in a chemical solution. This solvent removes about 99 per cent of the pure crude oil from the flake. After the oil has been removed, the oilseed flake is then cleaned, toasted and ground to improve its nutritional value. In stock feed manufacture, the disadvantage of this method is that there is need to set up an extra facility for toasting the cake (in order to eliminate anti-nutritional factors and improve nutritive value) before mixing it with other ingredients.

Equipment for vegetable oil solvent extraction is not available locally. However, the equipment can be imported as a complete set from Bühler AG[h], Troika[k] of India or HTTC[g] of China. Importing from Bühler[h] (South Africa) has an advantage in that it is nearer to Zimbabwe (proposed factory site in this case study) and it offers consultancy, installation, start-up and customer training services.

Mechanical extraction

The mechanical or expeller method of extracting oil is simpler than the solvent extraction. Seeds are fed into a mechanical press, which applies hydraulic force to expel the oil from the flake. The advantage of this method is that the heat generated during the process is enough to produce oilseed meal of optimal quality. However, oil extracted in this manner tends to have more impurities than that from the solvent extraction process. Physical extraction is also less efficient, resulting in up to 9–12 % less oil being extruded than when solvent extraction is employed. As a result, oilseed meal produced by this method contains a higher fat content (5 to 6% fat) than solvent-extracted meal (about 1%).

Local engineering firms have not been able to manufacture high capacity mechanical oil extruders. Intermediate Technology[d] manufactures a Chinese designed vegetable oil expresser that is capable of extracting between 20 and 25 litres of oil per hour from soya bean seed and between 10 and 12 litres per hour from cotton seed or sunflower seed. Renox[i], a firm in Harare, is the local distributor of the Hilax vegetable oil expressing machine. The biggest of them, the Hilax 105 is only capable of producing between 20 to 35 litres of oil per hour from soya bean, cotton seed and sunflower.

Larger capacity oil presses have to be imported. HTTC[g] of China manufactures the Helical oil press machine capable of extracting between 150 and 250 litres of oil per hour, among other smaller ones. The Alimentarmash Joint-Stock Company[j] of Mouldova manufactures a wide range of oil presses. One of them is the Line M8-MKA, with the capacity to produce between 416 and 625 kg of oil per hour from all oilseeds. At 27 tonnes and occupying 10 square metres, it is the heaviest and highest producer from the company. The GOBIND Expeller Co.[cc] of India manufactures the Anand (Super series) Oil Expeller. This machine produces between 25 and 30 tonnes of oil per day (24 hour) from sunflower, and between 20 and 25 tonnes from either cotton seed or soya bean.

Oil refinery and bottling

Crude oil obtained from solvent extraction and mechanical expellers should be purified before it can be marketed. Purification involves filtration and deodorizing the oil. Filtration removes the solid impurities that affect the texture (viscosity) of the product. Deodorization is intended for removing the specific smells from the vegetable oil and also reducing the acid number of the oil.

Large-scale vegetable oil purification equipment is not available locally. HTTC[g] supplies a complete set of equipment required for oil refinery. The Alimentarmash Joint-Stock Company[j] manufactures the M8-KFM vacuum filtration complex with a capacity to filter up to 170 kg of oil per hour.

This could be purchased together with the M8-LDM line for deodorizing the vegetable oil.

The Alimentarmash Joint-Stock Company[j] also supplies the M8-MRSH machine for bottling and corking. It is intended for dosage pouring out, using cylindrical bottles (polyethylene and glass) with the capacity from 0.5 up to 2 litres. The diameter of the bottles should be 60–100 mm, and the height should be 150–340 mm. Corking of the bottles is with screw covers. The M8-MRSH processes 300 bottles per hour.

Local firms like DPC[w] and Mukundi[x] manufacture bottles and containers in wide range of plastic material. Zimglass[y] manufactures glass bottles.

Coolers/driers

Coolers and driers are required in order to speed up the cooling/drying process in mechanically extracted oilseed flakes as well as in feed ejected from cubers or blenders. Cooling the dry feed coming from blenders is vital before vitamins can be added in. Vitamins are heat-labile, and are denatured if they are added to hot feed. There are no coolers/driers specifically designed for the stock feed manufacturing industry available locally. However, consultation with some local ventilation engineering firms like Ozone Engineering[n] and Fanquip[o] could result in the design of appropriate cooling/drying systems on-site. Engineer John Cooper of Ozone[n] has designed a fluidised bed cooling system for one of the country's leading players in the stock feed industry. Ozone Engineering[n] also have in stock a rotary drier/cooler that needs some minor refurbishment. The rotary drier/cooler can also double up as a feed mixer depending on the structure of the manufacturing process at a specific stock feed plant. Elsewhere, Bühler[h] supplies counter current coolers and horizontal coolers meant for stock feed plants.

Weighing

Different weighing machines are required for weighing micro and macro ingredients. Scales of different capacities are available locally, from Avery Berkel[p]. For a stock feed processing plant, where precision is of paramount importance, digital scales are highly recommended. Avery Berkel[p] supplies digital scales capable of measuring weights between 500 g and 300 kg, with the screen display mounted on the wall for convenient reading. Another industrial scale supplier that can be approached locally is the M & M Scale Company[q]. Bühler[h] supplies micro-differential dosing scales for minor ingredients and the Tubex electric weighing system for macro ingredients.

Feed bins

After weighing and proportioning, the ingredients are pooled into 'feed bins'. These are containers used for transferring various feed mixes to the mixers. If the transfer of the feed mixes in the bins is manually operated, then the feed bins might not necessarily be special ones. Some local plastic moulders like Storatank[r] could be approached in order to have some bins tailor-made. If an automated system is opted for, then firms like Bühler[h] could be approached.

Feed mixers

Feed mixers assist in accomplishing homogeneity in the feed. Bulk ingredients like maize, oilseed cakes and hay are mixed first until homogenized. Micronutrients like salt, limestone flour, minerals and vitamins are then added last and the mixer is run again. Apart from the rotary drier/cooler from Ozone[n], there are no known sources of feed mixers in Zimbabwe. Bühler[h] supplies two kinds of mixers. The Speedmix batch mixer is used to mix smaller batches of ingredients. The homogenizer mixes large batches of all the necessary ingredients to form a uniform product.

Block mould and press

Feed mixtures meant for making blocks are transferred into block moulds. The feed is then compacted from the top by a high-pressure hydraulic press. For a complete set of moulds and press, Bühler[h] South Africa designs and installs for customers in the Southern African region.

Cubers and pelletizers

In order to make cubes and pellets, feed mixtures are transferred into cubers and pelletizers, respectively. These machines are not available locally. The nearest source in the region is Bühler[h] South Africa. It supplies a complete set, which includes Kubex pellet mills, hydraulic roll adjustment device, pellet mill controller and the pellet crumbler. An oscillating sieve or shaker is also required to separate pellets from crumbles.

Bagging

A number of bagging and closing systems are available from the external market. Bühler[h] South Africa installs various bagging systems, including a gross bagging scale, an automatic bag closing unit and a stitching conveyor belt. The Model 6CM automatic bagging machine from American-Newlong Inc[s]. is capable of filling up to 1800 bags per hour. In order to source packaging material for the feed, there are a number of bag and sack manufacturers in Zimbabwe. Highfield Bag[m], Natpak[t], Saltrama[u] and Zimbabwe Grain Bag[v], are some of the local suppliers of labelled bags.

Dust extraction

Enclosures within the stock feed processing plant tend to accumulate a lot of dust. Locally, Tanroy Engineering[f] has designed and installed a dust extractor in the bagging room of one of the emerging players in stock feed manufacture in Zimbabwe. Bühler[h] installs cyclone separators with super-jet filters and fans to perform the same task.

In-house laboratory

As part of the processing factory's quality control program, there is need to set up an in-house laboratory. Quality control standards in the stock feed manufacturing industry include proximate analysis and testing for anti-nutritional factors like trypsine inhibitors (plus urease activity), gossypol, rancidity and salmonella. Proximate analysis of the feed involves determining the dry matter (DM), organic matter (OM), crude protein (CP), fibre (crude fibre, neutral detergent fibre or acid detergent fibre) and fat or ether extract (EE) levels in feed samples. In order to have stock feed products of acceptable standards, the methods used for laboratory analyses should be internationally recognized. It is therefore recommended that the in-house laboratory staff should have access to the Association of Official Analytical Chemists (AOAC) Handbook[Ψ]. Latter versions of the handbook are available. Procedures of the AOAC are widely recommended in Zimbabwe and internationally. The handbook also lists the laboratory equipment required for the various tests.

There is no known source of the handbook locally or regionally. However, copies of the handbook can be ordered online directly, via the AOAC website[z]. In order to familiarize with the type of equipment required and access to procedures used, the in-house laboratory could establish linkages with AREX's Chemistry and Soils Research Institute[aa] or the University of Zimbabwe's Department of Animal Science[bb]. These two institutions have vast experience in proximate analysis of animal feeds.

9 Conclusion

This case study was adapted from a consultancy report the author did for a Zimbabwean organisation. The following directory can be of help to would-be new stock feed

---

[Ψ] *Official Methods of Analysis, volumes* 1 & 2, fifteenth Edition, Virginia AOAC Inc., 1990

manufacturers, big or small. The letters a,b,c et cetera link the equipment supply to the stock feed processing scheme given above.

Equipment supplier directory

a. Jerylor (Pvt) Ltd. t/a Taylormade, 23 Watts Rd, Cnr S. Mazorodze/Leyland Rds, New Ardbennie, Harare. Tel: 666 461, 669244, 662 536. Fax: 669244. E-mail: Jerylor@samara.co.zw
b. DICHWE Implements (Pvt) Ltd, 1 Charles Prince Rd, Mt Hampden, Harare. Tel: 336131, 334 865, 336 962, 336 010.
c. Precision Grinders Engineers (Pvt) Ltd, 55 Craster Rd, Southerton, Harare.
Tel: 665631 – 5. Fax: 668628. E-mail: grinders@mweb.co.zw
d. Intermediate Technology Business Shop, 216 Willowvale Rd, Harare. Tel: 669773/4. Fax: 669773.
E-mail: itbshop@icon.co.zw; itbshop@africaonline.co.zw
e. Product (Pvt) Ltd, 84 Kelvin Rd South, Graniteside, Harare. Tel: 753243, 754192, 755120, 755224, 755082/3 755743. Fax: 757224.
E-mail: product@africaonline.co.zw
f. Tanroy Engineering (Pvt) Ltd, 179 Loreley Cres, Msasa, Harare. Tel 487791 – 3. Fax: 487794.
E-mail: tanroy@africaonline.co.zw
g. HTTC, Bldg 11, No.6 Nongye Raod, Zhengzhou city, He, 450002, China. Telephone: 0086-371-3812887.
E-mail: shang5@371.net
h. Bühler AG (Pty), Eva Park, Block A, 2nd Floor, Cnr D F Malan Dr/Judges Ave, Crosta, Randburg, 2194. PO Box 551, Crosta, 2118, RSA. Tel: +27 11 380 8000 Fax: + 27 11 380 8010
E-mail: buhlersa@global.co.za
i. Renox (Pvt) Ltd, 61 Prices Road, Mt Pleasant, Harare. Tel: 332574. Fax: 335097
j. Alimentarmash Joint-Stock Company, 12 Kishinau Street, Meshterul Manole, MD2044, Republic of Mouldova. Phones and faxes: 47-12-25, 47-10-96 marketing department ;
47-13-36 office; 47-43-60 sales department; Chisinau area code (3732). E-mail: almash@mdl.net
k. TROIKA, 6th Floor, Embassy Centre, Narim Point, Mumbai 400, 021 India. Tel: 00-9-(22)-2834429, 2834334, 2834515. Fax: 00-91-(22)-2823778.
E-mail: troika@vsnl.com

m. Highfield Bag (Pvt) Ltd, Cnr Douglas/Canberra Rds, Workington, Harare.
Tel: 620691 – 9, 620695 – 7. Fax: 620690.
n. Ozone Engineering (Pvt) Ltd, 21 George Ave, Msasa, Harare. Tel: 486307, 486141, 480622, 480623. Fax: 486307.
o. Fanquip (A division of Farmquip Pvt. Ltd.), 14 Whitesway, Msasa, Harare. Tel: 487813–6, 487740, 480544-6. Fax: 487743.
E-mail: farmquip@mail.pci.co.zw
p. Avery Berkel (Pvt) Ltd., 4 Conald Rd, Graniteside, Harare. Tel: 758492-9. Fax: 758490/91.
E-mail: mcavery@africaonline.co.zw; averyit@africaonline.co.zw
q. M & M Scale Company, 114 Harare Street, Harare. Tel: 723599, 721757, 011 411 831, 011 411 832. Fax: 721757
r. Storatank (Pvt) Ltd., 10–157 Citroen Rd, Msasa. Tel: 486411, 486396, 011 608 458/9 Fax: 486930
s. American-Newlong Inc., 5310 S. Harding Street, Indianapolis IN 46217, United States
Tel: (317) 787-9421. Fax: 786-5225.
E-mail: newlong@american-newlong.com
t. Natpak Zimbabwe (Pvt) Ltd, Kelso Rd (Off Stirling Road), Workington, Harare. Tel: 748041 – 3. Fax: 753738
u. Saltrama Plastex, 110 Lytton Rd, Workington, Harare. Tel: 621840 – 9. Fax: 621836.
E-mail: saltrama@africaonline.co.zw
v. Zimbabwe Grain Bag (Pvt) Ltd, 11 Dunlop Rd, Donnington, Bulawayo. Tel: 77574 – 6, 76776. Fax: 76779.
w. DPC (Die & Pressure Castings), 7 Martin Drive, Msasa, Harare. Tel: 498951–5486407, 486496. Fax: 487258.
E-mail: dpccolin@id.co.zw
x. Mukundi Plastics, 110 Lytton Rd, Workington, Harare. Tel: 621850 – 8. Fax: 621859.
E-mail: mukundi@africaonline.co.zw
y. Zimbabwe Glass Industries (Zimglass) Ltd., Portland Rd, Heavy Ind. Site, Gweru. Tel: 22801 – 5. Telex: 77709. Fax: 23512 or 21981. E-mail: zimglass@samara.co.zw
z. Association of Official Analytical Chemists (AOAC) Inc., 481 North Frederick Avenue Suite 500, Gaithersburg, Maryland 20877-2417, USA. Tel: +1-301-924-7077 Fax: +1-301-924-7089
E-mail: aoac@aoac.org Website: http://www.aoac.org
aa. Chemistry and Soils Research Institute. Agricultural Research Centre, AREX, 5th Street Extension. Harare. Tel: 704531, 704541.
bb. Department of Animal Science, University of Zimbabwe, Mount Pleasant, Harare. Tel: 303211 ext 1409

cc. GOBIND Expeller Co. 645 Industrial Area – B, Ludhiana 141003, India. Tel: 91-161-2531591, 2530711. Fax: 91-161-2537973, 2427576. E-mail: gobindindia@glide.net.in